ENERGY POLICIES, POLITICS AND PRICES

SOLID-STATE AND COMPACT FLUORESCENT LIGHTING

LESSONS LEARNED ON THE WAY TO MARKET

ENERGY POLICIES, POLITICS AND PRICES

Additional books in this series can be found on Nova's website under the Series tab.

Additional e-books in this series can be found on Nova's website under the e-book tab.

ENERGY POLICIES, POLITICS AND PRICES

SOLID-STATE AND COMPACT FLUORESCENT LIGHTING

LESSONS LEARNED ON THE WAY TO MARKET

JOAN L. THOMPSON
EDITOR

nova publishers
New York

Library of Congress Cataloging-in-Publication Data

ISBN: 978-1-63482-881-9

Published by Nova Science Publishers, Inc. † New York

CONTENTS

PREFACE

The purpose of this book is to document early challenges and lessons learned in the SSL market development as part of the DOE's SSL Program efforts to continually evaluate market progress in this area. This book summarizes early actions taken by DOE and others to avoid potential problems anticipated based on lessons learned from the market introduction of compact fluorescent lamps and identifies issues, challenges, and new lessons that have been learned in the early stages of the SSL market introduction. Furthermore, this book reviews efforts to increase market acceptance of compact fluorescent lamps (CFLs) and barriers to that acceptance.

Chapter 1 – In recent years, solid-state lighting (SSL) has emerged as a promising new lighting technology that could fundamentally alter and improve lighting systems and significantly lower energy use and costs. However, SSL's full performance and energy savings potential is far from realized or assured. The U.S. Department of Energy (DOE) has invested public funds in research and development to support advancements in the performance and energy efficiency of SSL technology, as well as a range of activities intended to increase the likelihood of rapid market uptake of new SSL products.

The purpose of this report is to document early challenges and lessons learned in the SSL market development as part of the DOE's SSL Program efforts to continually evaluate market progress in this area. This report summarizes early actions taken by DOE and others to avoid potential problems anticipated based on lessons learned from the market introduction of compact fluorescent lamps and identifies issues, challenges, and new lessons that have been learned in the early stages of the SSL market introduction.

Chapter 2 – Compact fluorescent lamps (CFLs) were introduced in the 1970s; however, it has taken more than 30 years for them to gain widespread recognition in the U.S. residential lighting market. Government and utilities have invested significant public and ratepayer funds in promoting the technology because CFLs represent an effective energy efficienct resource: they decrease lighting energy consumption by one-half or more for each incandescent lamp replaced. While CFLs still have only about 2% of the national market in terms of unit sales, they have achieved higher market penetration in some parts of the country, including California (5%) [43] and the Pacific Northwest (12%) [44].

This report reviews efforts to increase market acceptance of CFLs and barriers to that acceptance. The report was prepared by the Pacific Northwest National Laboratory for the U.S. Department of Energy's Office of Building Technologies, Emerging Technologies Program. Lessons to be learned from this study of CFLs are identified in hopes of assisting future market introduction efforts for other promising energy-efficient lighting technologies.

In: Solid-State and Compact Fluorescent Lighting ISBN: 978-1-63482-881-9
Editor: Joan L. Thompson

Chapter 1

SOLID-STATE LIGHTING: EARLY LESSONS LEARNED ON THE WAY TO MARKET*

L. J. Sandahl, K. A. Cort and K. L. Gordon

EXECUTIVE SUMMARY

In recent years, solid-state lighting (SSL) has emerged as a promising new lighting technology that could fundamentally alter and improve lighting systems and significantly lower energy use and costs. However, SSL's full performance and energy savings potential is far from realized or assured. The U.S. Department of Energy (DOE) has invested public funds in research and development to support advancements in the performance and energy efficiency of SSL technology, as well as a range of activities intended to increase the likelihood of rapid market uptake of new SSL products.

The purpose of this report is to document early challenges and lessons learned in the SSL market development as part of the DOE's SSL Program efforts to continually evaluate market progress in this area. This report summarizes early actions taken by DOE and others to avoid potential problems anticipated based on lessons learned from the market introduction of compact fluorescent lamps and identifies issues, challenges, and new lessons that have been learned in the early stages of the SSL market introduction.

---

* This is an edited, reformatted and augmented version of a January 2014 report, PNNL-23059, prepared under contract by Pacific Northwest National Laboratory for the U.S. Department of Energy.

Actions by DOE, voluntary energy-efficiency programs, and standards organizations have helped the U.S. market to avoid some problems with early SSL products. Standardized testing, minimum performance and reporting requirements, and publication of testing and demonstration results have made it more difficult for poor-performing products to remain on the market, and rewarded manufacturers whose products perform well. As SSL technologies, or more specifically light-emitting diode (LED) technology, continue to develop and reach into most general illumination applications, the steep learning curve that lighting industry and all stakeholders continue to climb will offer many more lessons to learn.

This study identifies and characterizes 12 key lessons that have been distilled from DOE's SSL Program results. These key lessons include the following:

- Lesson 1: Rigorous testing requirements adopted in the early days of SSL industry development were necessary to counter exaggerated claims of performance by some manufacturers, but they eventually led to unreasonably high testing costs
- Lesson 2: Despite the promise of long life, there is no standard way to rate the lifetime and reliability of LED products
- Lesson 3: Specifiers prefer complete families of products, but the rapid evolution of LED technology presents a challenge to manufacturers in creating and maintaining complete product lines
- Lesson 4: The range of color quality available with LED-based products and the limitations of existing color metrics may confuse users
- Lesson 5: The color delivered by some LEDs shifts over time, enough to negatively impact adoption in some applications
- Lesson 6: Some LEDs flicker noticeably, which may negatively impact adoption in some applications
- Lesson 7: LEDs can cause glare, which may negatively impact adoption in some applications
- Lesson 8: Achieving high-quality dimming performance with LED lamps is difficult, but improving
- Lesson 9: Greater interoperability of lighting control components and more sensible specifications of lighting control systems are required to maximize the energy savings delivered by LED-based sources
- Lesson 10: Lack of LED product serviceability and interchangeability has created market adoption barriers in certain sectors

- Lesson 11: Existing lighting infrastructure limits the full potential of SSL; more effort is needed to open the doors to new lighting systems and form factors
- Lesson 12: Programs that provide ways to identify quality LED products have helped support market adoption

Many additional lessons could be listed, but this report focuses on areas where ongoing challenges exist and/or useful information can be applied going forward. Some of the lessons learned from problems that have been addressed or superseded by technology are discussed in Section 2 of this report.

ACRONYMS AND ABBREVIATIONS

ANSLG	American National Standard Lighting Group
BUG	backlight, uplight, and glare
CALiPER	Commercially Available LED Product Evaluation and Reporting
CCR	constant current reduction
CCT	correlated color temperature
CFL	compact fluorescent lamp
CIE	International Commission on Illumination
CRI	color rendering index
DLC	Design Lights Consortium
DOE	U.S. Department of Energy
HID	high-intensity discharge
HPS	high-pressure sodium
IALD	International Association of Lighting Designers
IES	Illuminating Engineering Society
LED	light-emitting diode
MSSLC	Municipal Solid-State Street Lighting Consortium
NEEP	Northeast Energy Efficiency Partnerships
NEMA	National Electrical Manufacturer's Association
NGL	Next Generation Luminaires
NGLIA	Next Generation Lighting Industry Alliance
PWM	pulse-width modulation
QPL	Qualified Products List
R&D	research and development
SPD	spectral power distribution

SSL	solid-state lighting
TINSSL	Technical Information Network for SSL
UGR	Uniform Glare Rating
VCP	Visual Comfort Probability

1.0. Introduction and Scope

In recent years, solid-state lighting [1] (SSL) has emerged as a promising new lighting technology that could fundamentally alter and improve lighting systems and significantly lower energy use and costs. SSL is controllable and directional, with the potential to change the way we light buildings and outdoor areas by putting light where it is needed, when it is needed, while eliminating wasted light and drawing just a fraction of the power used by traditional light sources. However, SSL's full performance and energy savings potential is far from realized or assured. The technology, market, and infrastructure development needed to capture this potential is significant and likely to take years.

The U.S. Department of Energy (DOE) has invested public funds in research and development to support advancements in the performance and energy efficiency of SSL technology, as well as a range of activities intended to increase the likelihood of rapid market uptake of new SSL products. DOE's strategy in supporting development of SSL was informed significantly by lessons learned during the development and commercialization of another relatively recent lighting technology: compact fluorescent lamps (CFLs). Early CFLs had significant quality and market problems, as acknowledged by the lighting industry and documented by DOE. [2] CFLs ultimately achieved substantial market share, providing huge gains in efficiency and longevity for household lighting, but CFLs never achieved widespread popularity and advances in CFL technology have plateaued. DOE applied lessons learned from the CFL market experience to the nascent light-emitting diode (LED) market, making objective product information available to the market and providing important feedback to the industry to address quality problems early and aggressively. This approach has served the development of LEDs well, as significant progress has been made in the performance of SSL and it is now apparent that LED technology [3] has merely begun to achieve its full market and energy savings potential, and replacement lamps are but one segment of the market for LED lighting. Larger potential market impacts are evident, for example, in outdoor area and roadway lighting, where LED replaces high-intensity discharge (HID) sources, and commercial interior lighting, where LED replaces a number of sources including pin-based CFLs and, increasingly, linear fluorescent sources.

DOE has set aggressive and ambitious goals for SSL research and development (R&D). DOE's price and efficacy projections and goals are listed on Table 1.1. LED performance is improving so rapidly that DOE has revised its projections upward three times in the past 5 years. The current goal is to achieve a 266 lumen/watt efficacy at $0.50 per thousand lumens by 2030 (see Table 1.1). [4]

Table 1.1. Summary of LED Package Price and Performance Projections

	2012	2013	2015	2020	Goal
Cool-White Efficacy (lm/W)	150	164	190	235	266
Cool-White Price ($/klm)	6	4	2	0.7	0.5
Warm-White Efficacy (lm/W)	113	129	162	224	266
Warm-White Price ($/klm)	7.9	5.1	2.3	0.7	0.5

Now that LED lighting is available for most lighting applications and several years of market and installation experience have accumulated, DOE is assessing early lessons and observations from the LED market thus far. This report identifies some of the early actions taken in support of LED market introduction, and focuses on current challenges, lessons learned, and their implications for the future. These lessons are primarily based on insights from DOE's Market Development Support Program, which began in 2006 and serves as an independent third party providing LED technical information, performance data, education materials, and other support for market development and adoption of high-quality, high efficiency LED lighting. DOE SSL Program activities involve coordination with a wide variety of stakeholder groups including manufacturers, utilities and efficiency groups, designers, end users, and distributors.

The scope of this report focuses on LED lighting used for white light, general illumination applications, including interior and exterior lighting of buildings and facilities, and exterior lighting including street, roadway, parking lot, parking garage, and other forms of outdoor lighting. Vehicle, traffic, video display, screen, indicator, and monochromatic lighting applications are outside the scope of this report. Technical performance issues directly related to market development and adoption of LED general illumination products are addressed in the report, as are market barriers, standards, lighting quality, lighting design, and issues related to lighting energy-efficiency programs.

Section 2 discusses key lessons from CFL market introduction that informed early strategies implemented by DOE and others in the LED market. Section 3 identifies and characterizes 12 lessons and observations from the LED market introduction experience to date, focusing on lessons for which ongoing challenges exist and/or useful information can be applied going forward, along with implications for the future. Section 4 provides conclusions.

2.0. CFL Lessons Applied to Early LED Market

The Energy Policy Act of 2005 established the Next Generation Lighting Initiative, directing DOE to "support research, development, demonstration, and commercial application activities related to advanced solid-state lighting technologies based on white light-emitting diodes" (Section 912, (b)). DOE had laid the groundwork with several technology roadmapping reports developed in cooperation with the SSL industry in the early 2000s. In 2006, looking ahead to imminent market introduction of the first LED-based general illumination products, DOE decided to draw on the collective experience of the lighting industry, energy-efficiency program sponsors, and the federal government in promoting CFLs as an energy-efficient alternative light source. DOE commissioned the report, *Compact Fluorescent Lighting in America: Lessons Learned on the Way to Market*, [5] to identify key lessons and assess their potential to inform DOE strategy in support of LED market introduction and development.

Summary of Key Lessons Learned from CFL Experience

Coordination and collaboration is key

- Coordinate and collaborate at a national level.

Establish standards and product testing

- Collaboratively establish minimum performance requirements.
- Back up long-life claims with standard-based projections and/or guarantees.

Introduce new lighting technology first in applications where benefits are clearly established

- Study market structure to see how best to introduce the technology.
- Know and admit technology limitations.

Respond to the market and resolve problems/issues quickly

- Determine and address compatibility issues with standard or conventional light fixtures and designs.
- Be aggressive about dealing with technology failures and issues that affect user satisfaction and/or the main benefit claims.
- Don't launch a product until performance issues are ironed out.

The 2006 CFL lessons learned study provides an analysis of the market introduction of CFLs, with an emphasis on identifying lessons that could be applied to the introduction of other new lighting technologies, such as SSL. In particular, the study explored reasons behind the slow acceptance of CFLs even though they last up to ten times longer than standard incandescent bulbs and use at least two-thirds less energy to provide the same amount of light. The study documented lessons that apply directly to LED market development (see text box for summary of CFL lessons learned), and many of the lessons learned from the study have guided DOE's planning and implementation of its SSL Program. The following summarizes the key lessons learned from CFL experience and how they have been applied in the early LED market experience.

2.1. Coordination and Collaboration Is Key

Coordination across the multiple actors and stakeholders in the lighting market was lacking at the time CFLs were introduced to the market. This slowed the ability to influence and address the quality problems that inhibited market adoption. Cognizant of this experience, a number of government, utility, and industry-supported collaborative efforts have been developed to help avoid potential problems and speed the market adoption of LEDs, including the following:

- Next Generation Lighting Industry Alliance (NGLIA) – Manufacturers came together in 2003 to form the NGLIA, an alliance of for-profit lighting manufacturers formed to accelerate SSL development and commercialization through government-industry partnership. DOE signed a Memorandum of Agreement with NGLIA in February 2005 detailing a strategy to enhance the manufacturing and commercialization focus of the DOE portfolio by utilizing the expertise of NGLIA. [6] NGLIA coordinates with DOE in project development, review of technical specifications, development of new standards and test procedures, and cosponsoring of DOE events.

- The Illuminating Engineering Society (IES) and International Association of Lighting Designers (IALD) – The IES signed a Memorandum of Understanding with DOE in July 2006 to strengthen their ongoing partnership and commitment to improve the efficient use of energy and to develop standards with a strong energy-efficiency focus. A Memorandum of Understanding between DOE and the IALD was signed in 2008 to strengthen coordination in support of efficient lighting systems and equipment between DOE and lighting designers. [7] Both IES and IALD are involved with DOE in sponsoring the Next Generation Luminaires design competition, review of technical specifications, development of new standards and test procedures, and cosponsoring of DOE events.
- Technical Information Network for SSL (TINSSL) – To better coordinate and share SSL information with key stakeholder groups, including regional energy-efficiency organizations, utilities, lighting trade groups, and other stakeholders, DOE established TINSSL in 2006. Members share information about the technical progress of LED technologies through monthly teleconferences, national events, and distribution of fact sheets, market studies, and technical reports. DOE also hosts an annual SSL market introduction workshop and other information exchanges that draw together experts from industry, academia, and research organizations.
- ENERGY STAR – ENERGY STAR qualifications became available for LED fixtures starting in 2008 and LED integral lamps in 2010. [8] Working with a wide range of stakeholders, the program develops product qualification criteria that cover primarily residential lighting applications. During 2012, CFL technology continued to account for the majority of lighting product promotions according to the U.S. Environmental Protection Agency's 2012 ENERGY STAR Summary of Lighting Programs, [9] but LED promotions were growing in number.

- Design Lights Consortium (DLC) – The DLC is an initiative of Northeast Energy Efficiency Partnerships (NEEP) started in 1998 to raise awareness of the benefits of efficient lighting in commercial buildings, in support of energy-efficiency programs implemented by NEEP member utilities in the Northeast. In 2010, utilities around the country joined the DLC to initiate the Qualified Products List (QPL) for SSL products meeting minimum performance criteria for commercial grade LED luminaires. DLC qualification has become a requirement for many utility and energy-efficiency programs across the U.S. and Canada.
- National and regional efforts by utility and energy-efficiency groups – Utility and efficiency programs, many of which coordinate at the national and regional levels, have invested significant resources into better understanding the energy savings potential of LEDs and the most appropriate early markets to address with incentive programs (many have programs well underway), and understanding market barriers and how to best support the market adoption of LEDs in various market sectors. These organizations include NEEP, Northwest Energy Efficiency Alliance, the Midwest Energy Efficiency Alliance, and the Consortium for Energy Efficiency. Individual utilities that are part of these organizations have also made great strides. For example, starting in 2008, three of California's investor-owned utilities (Southern California Edison, Pacific Gas and Electric, and San Diego Gas and Electric) undertook a series of market transformation activities focused on LEDs. These market transformation activities are included in an annual report on the topic [10] and include both residential and commercial-focused programs targeting (1) the technical advancement of LEDs; (2) lighting education and information; and (3) codes, standards, policies, and program administration.

All of these collaborative efforts internalized lessons from the CFL experience, including attention to the accuracy of product performance claims, a commitment to product testing and performance verification, and attention to lighting quality as well as energy efficiency. National level cooperation was necessary to work effectively with the lighting industry and communicate consistent messages about product quality, performance, and efficiency. In some cases, utilities, regions, or individual states have developed their own requirements above and beyond national requirements, but all are underpinned by nationally coordinated test methods, voluntary labeling programs, technical information, and qualified product lists.

2.2. Establish Standards and Product Testing

To help establish trust in the market for LEDs, lighting organizations and DOE began work on LED standards and test procedures soon after DOE initiated its SSL Program. The unique nature of LEDs meant that existing test procedures to measure and evaluate traditional lighting sources could not all be applied to LED technology. As a result, the lighting industry and existing standards bodies, with support from DOE, developed new standards and test procedures for LEDs, including:

- ANSI C78.377-2011 – *Specifications for Chromaticity of Solid State Lighting Products*; originally published in February 2008, updated in 2011, and available for download from the ANSI website.
- IESNA LM-79-08 – *Method for Electrical and Photometric Measurement of SSL Products*; in revision, current version is 2008.
- IESNA LM-80-08 – *Method for Measuring Lumen Maintenance of LED Light Sources*; in revision, current version is 2008.
- IES TM-21-11 – *Method for Projecting Long Term Lumen Maintenance of LED Light Sources*; published in 2011.

DOE started the Commercially Available LED Product Evaluation and Reporting (CALiPER) program in 2006, driven by the lack of reliable LED product performance information in the market at that time. CALiPER purchased LED products through normal market channels (at retail, through distribution, online), tested them, published reports detailing the products' tested performance compared to manufacturer claims, and compared LED product performance to conventional lighting technologies. At the beginning, the standard test methods for measuring LED product performance were in development but not yet available. DOE worked with independent testing laboratories to test early LED products, based on the draft LM-79 photometric test procedure under development by the IES. This collaboration among DOE, testing laboratories, and the IES test procedures committee provided data and feedback loops that helped accelerate the completion of the LM-79 test procedure.

LM-79 and ANSI C78.377 were critical early standards that helped bring order to the chaotic LED market, providing a common basis for evaluating and comparing the performance of LED products to one another and to conventional lighting technologies. Nearly all national, regional, state, and utility energy-efficiency programs reference these industry standards.

2.3. Introduce New Lighting Technology First in Applications Where Benefits Are Clearly Established

While LED products were marketed for many different lighting applications almost from the beginning, the lighting industry largely focused on applications that were most ripe for market adoption, both in terms of performance and cost. Outdoor street and area lighting, for example, where long operating hours and attractive maintenance savings outweighed the challenges of the relatively high required light outputs, were early successful applications for LED lighting, and continue to advance quickly. Recessed downlights took advantage of LED directional output and the availability of the fixture housing for thermal management.

LED replacements for directional lamps such as MR16s and PARs [11] have also proliferated in the market, but GATEWAY [12] evaluations have identified performance problems in some installations due to overheating and optical control challenges. Similarly, CALiPER testing of early LED T8 replacements revealed low light output, poor light distribution, and minimal energy savings compared to high-performance fluorescent T8 lamps. [13] This category of products has since improved in efficacy and output, but light distribution and ease of installation issues continue.

To build on early success in LED streetlighting, DOE established the Municipal Solid-State Street Lighting Consortium (MSSLC) in 2010. The purpose of the Consortium is to share technical information and experiences related to LED street and area lighting demonstrations and to serve as an objective resource for evaluating new products on the market intended for those applications. Cities, power providers, and others who invest in street and area lighting were invited to join the MSSLC and share their experiences with an aim toward building a repository of valuable field experience and data. DOE's goal was to significantly accelerate the learning curve for buying and implementing high-quality, energy-efficient LED lighting. Response to the MSSLC has been rapid and enthusiastic. More than 375 members from 46 states are represented in the MSSLC.

LED products appear poised to capture a majority share of the streetlighting market over the next decade. The economic and energy cases are clear, but many of the nation's streetlighting owners are still in the early stages of understanding this new technology and, for example, how to distinguish appropriately performing products from others that are not. To help members successfully implement LED streetlighting projects, the MSSLC developed model specification documents, which members can use as a template and adapt to their own conditions. The MSSLC developed a similar document for adaptive control and remote monitoring systems for LED streetlighting. [14] Both model specifications will be updated regularly to keep pace with LED and controls technology development.

2.4. Respond to the Market and Resolve Problems/Issues Quickly

Addressing product limitations and factors that can cause user dissatisfaction is critical to avoid a lasting bad impression of the new technology. With this lesson from CFLs in mind, DOE took early action to evaluate and test LED product claims. In the first two rounds of CALiPER testing in 2007, less than 15% of the products had light output claims within 30% of tested values. [15] The publication of these discrepancies had an immediate impact: by the third round of CALiPER testing, the proportion of products with reasonably accurate light output claims (within 10% of tested values) had risen to 30%. [16] Of products CALiPER tested in 2012, about 67% had output claims within 10% of the tested value. [17]

DOE continues to use CALiPER testing, GATEWAY demonstrations, annual workshops, the DOE SSL website (www.ssl.energy.gov), fact sheets, R&D updates, studies, design competitions, and other methods to characterize and report the state of the technology and the market. The early and earnest involvement of industry in standards development and collaborative efforts through NGLIA, independent product testing through CALiPER, product qualifications by ENERGY STAR and DLC, and product verification by LED Lighting Facts [18] are all strong motivators to ensure accuracy in performance claims and help keep poor quality products off the shelves.

One example of a technical performance issue that has received early attention and efforts at resolution is compatibility of LED products with existing dimmers. Incompatibilities arise with the use of LED lighting products, which are powered by electronic drivers, on dimmers designed for incandescent loads. To date, there is no standard dimming interface, measurement protocol, or compatibility rating system that will ensure LED dimming performance with existing installed dimmers. (There is a dimming standard—NEMA SSL-7a—that addresses dimming compatibility between new lamps and dimmers that comply with the standard.) Manufacturers have taken different approaches to this problem. Most at least indicate in product packaging and literature whether the product may be used on a dimmer. Others publish lists of compatible dimmers. Advances in LED drivers have improved the chances for successful dimming of LED products, but compatibility and dimming performance problems are expected to continue for some time, as the existing installed stock of dimmers is replaced with dimmers designed for LEDs (see Section 3.8 for more information on this topic).

Another example where product concerns were addressed early relates to the environmental impact of light sources. The mercury content of CFLs became a significant concern in the lighting market, and sensitized users and researchers to possible similar issues with LED lighting products. To help address this concern, DOE commissioned a three-part study to investigate the entire life-cycle of selected LED lamps, from manufacturing and transport to use and disposal. [19, 20, 21] These reports demonstrated that, due to their energy efficiency, the impact of LED lamps over their entire life-cycle is significantly less than incandescent lamps, currently on par with CFLs, and expected to be less than CFLs with ongoing improvements in the efficiency of LED products. Part 3 of the study found that regulated elements in LED products were below restrictions set by the Federal Government, while all three tested light types (LED, CFL, and incandescent) exceeded at least one restriction imposed by the State of California, typically for copper, zinc, antimony, or nickel. LED products were found to be similar to cell phones and other types of electronic devices and should therefore be included in electronic waste recycling and disposal programs.

3.0. CURRENT LED LESSONS LEARNED

Actions by DOE, energy-efficiency programs, standards organizations, the SSL industry, and others—many in response to the CFL lessons learned discussed in the previous section—have helped the U.S. market to avoid some problems with early LED products. Standardized testing, minimum performance and reporting requirements, and publication of testing and demonstration results have made it more difficult for poor-performing products to remain on the market, and rewarded manufacturers who meet quality standards. As LED technology has continued to develop and reach into most general illumination applications, the lighting industry and all stakeholders have had to climb a steep learning curve, and much of the curve remains still ahead. LED products currently represent just over 4% of total installed lighting, [22] but their market share is growing rapidly. Now is a good time to pause and consider what has been learned thus far in LED general illumination market development, and the implications those lessons hold for the future.

The following sections present 12 key lessons learned in the development of LED as a general illumination light source. Many more lessons could potentially be listed, but the focus here is on areas in which ongoing challenges exist and/or useful information can be applied going forward. Lessons about problems that have largely been addressed or superseded by technology are not included. For example, two years ago, this lesson may have been salient: "LED replacements for 60-watt incandescent bulbs often do not provide adequate light output, light distribution, and performance." As of 2013, this problem has largely been addressed through improved LED output and efficacy, and product design improvements.

The 12 lessons that follow have been distilled from DOE SSL Program results, including

- CALiPER testing and exploratory studies
- performance data verification by LED Lighting Facts
- GATEWAY demonstration projects
- the multi-year R&D planning process
- regularly updated market studies and technology roadmapping efforts
- DOE annual workshops and major industry events
- extensive interaction with manufacturers, lighting designers and specifiers, retailers, building owners, municipalities, and utilities and energy-efficiency programs.

DOE recognizes the importance of aligning its R&D portfolio with industry partners, research and academic organizations, national laboratories, utilities and efficiency groups, designers, end users and distributors. The early successes of LED development and deployment are partly the result of these successful collaborative efforts. Likewise, each of these stakeholder groups and partnerships has a part in addressing the early LED market lessons outlined in Sections 3.1 through 3.12 of this report.

3.1. Lesson 1: Rigorous Testing Requirements Adopted in the Early Days of SSL Industry Development Were Necessary to Counter Exaggerated Claims of Performance by Some Manufacturers, but They Eventually Led to Unreasonably High Testing Costs

LED products face a higher testing burden than conventional lighting technologies due to a number of factors. Among the key factors that contributed to this high testing burden:

1) Characteristics of LED technology—especially the effect of thermal and electrical design on light output and efficacy—required a new test method (LM-79-08) based on absolute photometry, [23] instead of traditional relative photometry, for measuring LED luminaires. LED product performance could not be assumed to scale predictably with light source output or other product variations as is assumed with relative photometry, so the IES committee that developed LM-79 chose to base the new test procedure on absolute photometry.
2) Early CALiPER testing revealed the light output claims on many products were inaccurate, sometimes by very large margins. Once the LM-79 photometric test procedure became available, DOE encouraged buyers, specifiers, and energy efficiency programs to ask for test results for each product, creating a widespread expectation among LED buyers that all products should have an LM-79 report available. [24]
3) Energy-efficiency programs around the country adopted policies that required LM-79 reports to be available for each and every qualifying product, or adopted restrictive product family grouping policies that sharply limited the extent to which the LM-79 test on a product within a grouping could be used to qualify other products in that grouping.

As the LED industry has matured and many manufacturers have developed methods to accurately calculate product performance within a product family, DOE and efficiency program providers have taken steps to decrease this testing burden for manufacturers.

3.1.1. Significance

High testing burdens boost product prices, and slow down product introductions, which in turn can slow the rate of SSL market adoption and associated energy savings. Achieving a balance between the testing burden and the need for accurate performance data is important to encourage greater manufacturer participation in voluntary energy-efficiency and reporting programs, help level the playing field between LED and conventional lighting products, and bring more products and more innovation to the market quicker, while still providing the lighting market the necessary performance data.

3.1.2. Background

Absolute photometry and the resulting metric, luminaire efficacy, is a departure from traditional relative photometry used by the lighting industry. Traditionally, light sources were assumed to be omnidirectional, uniform, and scalable. A manufacturer could conduct one test on a single fixture and simply alter the magnitude based on the light source because the distribution remained the same or was "close enough" for industry needs. With the introduction of SSL, its unique properties, directionality, and thermal considerations, the industry developed LM-79, and for the first time every luminaire variation had to be tested. This dramatically increased the volume and costs of testing. ENERGY STAR, the DLC, and LED Lighting Facts all require LM-79 test results for product listing, and each have various product family grouping policies to allow manufacturers to extend the test results of one product to others in the product family. Most utility incentive programs reference one or more of these qualification programs. Many retailers require vendors to have their LED products listed by LED Lighting Facts. This demand for testing resulted in greater accuracy in product performance claims and greater confidence in LED products, but with the growing number of products being offered in the market, testing costs have skyrocketed.

3.1.3. Challenges

With the exploding number of products and product variations entering the market, rapidly rising testing cost has become an important issue for manufacturers. Manufacturer comments expressing frustration with testing costs and requirements have been submitted during voluntary criteria/program review cycles [25] and shared in industry trade press and events. While per product testing played a critical role in early product quality and buyer confidence in LED technology, it is becoming a hindrance to product line expansion. To meet the needs of the lighting market, testing costs will have to decrease to enable faster and more efficient LED product line development.

A growing number of manufacturers have climbed the LED learning curve and are fully capable of accurately predicting product performance within a product family based on limited testing and internally developed methods for extrapolating those results to other products in a family. However, there is still no standardized means of performing these calculations. Buyers and other users of manufacturer performance claims not based directly on photometric testing are thus left not knowing whether they can trust product performance claims. Third-party listing and qualification programs have to keep up with an ever-increasing number of applications to help users choose high-quality products.

3.1.4. Implications for the Future

1) Family grouping policies have been the primary means used by energy-efficiency programs for addressing the testing burden issue. The DLC allows for families of LED products to be qualified based on a single photometric test, along with documentation relating family members to the tested products. ENERGY STAR has a similar policy, and the LED Lighting Facts program has recently implemented a somewhat different family grouping policy, backed with random verification testing paid by manufacturer participants in that program. The voluntary qualification and listing groups should consider working together to continue to reduce the testing burden on manufacturers, while maintaining the integrity of product data through random verification testing. This can be accomplished by increased coordination, data sharing, and uniform requirements across programs.
2) The lighting community should consider continuing to seek methods and strategies to reduce the overall testing burden on manufacturers, as confidence in LED performance continues to increase via evolving product performance expectations, documented long-term performance studies, and additional standards to address overall system reliability.
3) Product listing and qualification programs should ensure that reduced testing requirements do not reopen the door to inaccurate or exaggerated product performance claims. Inaccurate performance claims discovered through random testing should be corrected promptly by the manufacturer or the product removed from qualified lists.

3.2. Lesson 2: Despite the Promise of Long Life, There Is No Standard Way to Rate the Lifetime and Reliability of LED Products

Understanding and communicating how LED products fail and how long they can last is challenging. Predicting LED product life and reliability is complicated by various technical factors, and a single performance metric typically does not fit all lighting types and applications well. [26]

3.2.1. Significance

While LED-based products hold the potential to achieve lifetimes that meet or exceed traditional lighting technologies, the life and reliability of LED products on the market varies widely, and manufacturer claims can be misconstrued by users who do not fully understand why LED products fail or the difference between lifetime and reliability. [27] This uncertainty about lifetime can impede market adoption of LED products in certain sectors, especially those sectors that are particularly cost-sensitive.

3.2.2. Background

Long life has been billed as a key advantage of LED sources. For conventional lighting systems (incandescent, fluorescent, HID), failure most commonly results when a lamp “burns out.” In almost all cases, other system components (e.g., ballast or luminaire housing) last longer than the lamp, and have lifetimes that are not dependent on the lamp. As a result, it has been sufficient to consider only the lifetime of the lamp itself.

Unlike conventional lighting systems, LED systems do not necessarily have “lamps.” In contrast, they may include a number of integrated components as in fully integrated luminaires or integral-driver lamps, which means that LED system performance is more affected by interactions among system components. These interdependencies complicate both the testing and the rating of the LED system lifetime, as the failure of any LED system component (the electronics, thermal management, optics, wires, connectors, seals, etc.) can directly or indirectly lead to product failure. In addition, while some LEDs may fail catastrophically (i.e., burn out), others may degrade over time to the point where they stop producing an acceptable quantity or quality of light (i.e., exhibit parametric failure). [28] There is currently no standard or well-accepted method for predicting the useful life or reliability of an LED product.

3.2.3. Challenges

At present, manufacturers are designing and developing LED sources using a wide range of materials and approaches, fueled by the rapid improvement and evolution of multiple core technologies. Lifetime and reliability are often at odds with cost in the battle to achieve greater market deployment. The difficulty in developing a standard method for predicting the useful life or reliability of an LED product and the wide variation seen in the lifetime and reliability of market-available products might not be significant barriers to LED product adoption in and of themselves. However, their simultaneous occurrence is causing significant uncertainty.

Market actors have attempted to reduce user uncertainty about the lifetime and reliability of LED products in various ways, ranging from unsubstantiated marketing claims, to the use of standardized metrics, to a more recent focus on product warranties. All present challenges. Unsubstantiated claims take time and effort to uncover, and initially may be overlooked if presented by a known brand. Standardized metrics, like L_{70} (hours of operation during which light output remains above 70% of initial output) provide some means for apples-to-apples comparisons, and represent an improvement over the early days when unsubstantiated claims were commonplace. However, many users do not understand the limitations of metrics and overestimate their value. For example, the use of $L_{70,}$ as determined using IES LM-80 data [29] and IES TM-21 [30] calculations, only addresses lumen maintenance of the LED packages, and says nothing about color shift or the lifetime and reliability of the LED driver and other components.

Recent testing undertaken by the LED Systems Reliability Consortium, in cooperation with DOE, found a high level of reliability of LED luminaires tested under extreme stress conditions, with failures occurring in the drive circuitry rather than the LEDs. [31] These findings underscore the need for a systems-level perspective in evaluating LED reliability, including light sources, drivers, optics, and other components.

While warranty coverage is the most important assurance for many users, manufacturer warranty terms vary significantly—making comparisons difficult if not impossible—and most manufacturers find it impossible to absorb the financial commitment required to provide warranty coverage on par with the expected (long) lifetime of well-designed products.

3.2.4. Implications for the Future

1) Variability in lifetime and reliability of similar products needs to be reduced to improve market adoption of LED products.

2) Uncertainty about the lifetime and reliability of LED products is slowly but steadily being reduced, as efforts are made to standardize methods for predicting lifetime and reliability. Standards developers are working on new tools and metrics, some of which will still be limited in their usefulness. Nevertheless, these tools and metrics represent one of many efforts that will be necessary to address this challenge.
3) Some manufacturers are finding the financial wherewithal to provide longer (often 10-year) warranties for products serving key applications and market sectors. Such efforts represent a key step in overcoming uncertainty, whereby users develop trust, first with specific products or perhaps application-specific offerings from proven manufacturers, and then with the technology as a whole.
4) Innovative LED product designs and the expected increased use of lighting controls may further complicate the measurement and reporting of LED product lifetime and reliability in the near future. LED sources that can adjust their output and color characteristics, either automatically or in response to lighting controls, do not operate under a single set of conditions. Varying conditions can improve or degrade lifetime and reliability. All market actors—especially DOE and the research community, standards developers, and manufacturers—will need to stay diligent and work together to explore the effect of such features on lifetime and reliability performance and product variation, and educate the various user communities.

3.3. Lesson 3: Specifiers Prefer Complete Families of Products, but the Rapid Evolution of LED Technology Presents a Challenge to Manufacturers in Creating and Maintaining Complete Product Lines

Lighting professionals have traditionally used families of lamps and luminaires in buildings. Families of luminaires may include products with different light intensities, aperture size, beam angle, direction of light, or trim colors, for example, while maintaining a similar appearance. This allows luminaires used throughout a space or building to have a consistent look and feel, and can also standardize maintenance and spare parts. Similarly, LED replacement lamps may be interchangeable in a luminaire, but lamps that offer different lumen outputs and beam angles make it easier to provide just the right light level and visual effect in different size spaces, or when illuminating a range of objects. The rapid evolution of products within the LED industry has challenged manufacturers trying to develop full product lines. In some cases, manufacturers lack the time to develop full product ranges, or it is difficult to keep products consistent with one another when LED packages and drivers are changing so rapidly.

3.3.1. Significance

Without the option of product families, specifiers may have trouble using LED products across their project. Using a downlight from one manufacturer, a wallwasher from another, and an adjustable accent light from a third may result in three LED products with mismatched light color in a room, or an inability to match fixture finishes. This may lead some specifiers to conclude that it is safer to use older, familiar technology instead, delaying the use of LED products until better options become available. It may also delay the penetration of LEDs into higher-end projects that use lighting professionals. This issue primarily applies to the commercial market for families of both replacement lamps and complete luminaires, where designers need a variety of luminaires/lamps to achieve the project goals. It may also apply to residential customers as they try to replace existing lamps in their homes. Some energy savings are left on the table when a specifier is forced to use a higher output lamp or luminaire than is needed in a space, because of a lack of lower-output options.

3.3.2. Background

The development of LED replacement lamps exemplifies the issue. LED MR16 lamps were developed early on, with a lumen output comparable to a 20W halogen MR16. Soon, LED package improvements led to a 35W halogen equivalent. However, most manufacturers opted to sell only the higher output lamps, when specifiers would have liked to have both a 20W and a 35W equivalent available for different areas of a hotel, for example.

To cite a commercial example, recessed 2x2 luminaires are often used in areas where a 2x4 luminaire will not fit, or where the 2x4 luminaire emits more light than needed. The specifier needs both sizes of luminaires on the job. Early LED troffers were offered in 2x2 **or** 2x4 sizes, but seldom both by the same manufacturer, delaying the adoption of the technology.

3.3.3. Challenges

The rapid evolution of products within the LED industry has challenged manufacturers trying to develop full product lines. In some cases, manufacturers lack the time to develop full product ranges, or it is difficult to keep products consistent with one another when LED packages and drivers are changing so rapidly. In addition, some manufacturers are either not aware or not significantly incentivized in terms of profit margin to understand that specifiers need a range of products within a given family to meet commercial design needs and standards.

The cost of photometric testing of luminaire families presents additional challenges to providing complete families, as the testing costs can be prohibitive since nationally recognized testing laboratories must separately test luminaires with different drivers, different trim colors and configurations, and different lumen output options.

3.3.4. Implications for the Future

1) Manufacturers should consider providing a range of lamps or luminaires with similar appearance, but with different photometric performance.
2) When a product is superseded by a product with greater lumen output, manufacturers should consider keeping both in the product line.
3) Without the option of product families, some specifiers may delay the use of LED products until these options become available.
4) Manufacturers may find the development of complete product families too costly given the rapidly changing LED market. However, manufacturers who can provide product families may find increased LED product specifications in commercial applications.
5) Developing one or more forms of relative testing standards may help manufacturers manage the cost of testing.

3.4. Lesson 4: The Range of Color Quality Available with LED-Based Products and the Limitations of Existing Color Metrics May Confuse Users

Unlike incandescent or halogen lamps, where color quality is very similar from one product to another, LED lamps can exhibit a wide variety of color appearance and color rendering attributes, which may confuse users. The emergence of LED technology has increased awareness of the limitations of existing color quality metrics and spurred the development of new metrics, further complicating efforts to identify the level of color quality that would be acceptable to most users and ensure widespread adoption of the technology.

3.4.1. Significance

Meeting customer expectations is paramount to widespread adoption of LED technology. While the availability of products with a range of color qualities may be an asset in one sense, the variability also presents an impediment to unknowledgeable users who are seeking to replace familiar products. In some cases, this has led to purchased products not meeting expectations, even though acceptable products may have been available. In different ways, the concern over color quality variation is present for both consumers purchasing lamps for residential applications and professionals specifying lamps for commercial applications. In some specific applications, such as roadway lighting, color quality may be less of a concern, but the accuracy of color metrics is still important.

3.4.2. Background

LED products can be engineered to produce white light across the full range of correlated color temperatures (CCTs), typically from 2700K to 6500K—not to mention varying values of D_{uv} for any given CCT. [32] This is very different from the incandescent or halogen lamps that many LED products are intended to replace—which are almost all between 2700K and 3000K (and have very small D_{uv} values). The range in CCT for LED-based products also exceeds the typical range for high-pressure sodium (HPS) or metal-halide-based streetlights, and is similar to what is available for fluorescent lamps.

Given the efficacy advantage of converting less energy to long-wavelength red emission, higher-CCT LED products tend to be more prevalent than higher-CCT CFLs, and in general were marketed and specified more frequently in the early years of LED adoption compared to CFLs. This was partially because LED products needed to have the highest efficacy possible to compete with other technologies and to provide enough life-cycle cost savings to justify higher purchase prices. Rising efficacy and the more widespread availability of lower-CCT products have provided some relief for this issue. Nonetheless, for some users, LEDs have been stigmatized as providing blue light, despite the availability of products with a full range of color qualities.

Beyond color appearance, LED lamps have more variability in the color rendering capability than most other technologies. All incandescent lamps have a color rendering index (CRI) near 100, and CFLs—with color quality that disappointed many consumers [33] —almost all have a CRI in the low 80s. This makes for easier decisions for specifiers and purchasers. A similar case is true for HPS, metal halide, and fluorescent lamps, which are all predominantly available in one CRI range. In contrast, it is easy to tune the spectral power distribution (SPD) of an LED to produce a chosen CRI. Over time, the market for interior products has largely settled on CRIs in the low 80s—similar to fluorescent lamps—but there is ongoing debate as to whether this is appropriate, given the relative ease—albeit often with modest efficacy tradeoffs—of achieving a higher CRI. A small percentage of products already have a CRI in the 90s.

Reliance on existing metrics, like CCT and CRI, has posed another challenge for LED technology. CCT [34] and CRI [35] were developed approximately 50 years ago, when the types of available light sources were substantially different. The context and purpose of the development of CRI differs from how CRI is used today, as are the capabilities to repeatedly perform complex calculations and the general state of lighting calculation methodology. LED products can have substantially different SPDs compared with conventional light sources, which can stress the capability of the CRI metric. Additionally, LEDs can easily be engineered to have a wide variety of chromaticities, either near or away from the black body locus. [36]

3.4.3. Challenges

The challenge posed by the variety of color quality attributes for LED products is multifaceted, and includes performance and measurement concerns, as well as communication concerns. In terms of performance and measurement, product variation can mean that the light emitted by two different model lamps having the same nominal CCT does not look the same. Given other concerns with the present state of the LED market—such as the lack of complete product families from some manufacturers—this can make it difficult to specify varying types of LED products for the same space. To address this concern, the American National Standard Lighting Group (ANSLG) and National Electrical Manufacturer's Association (NEMA) established standard C78.377-2008 (subsequently updated in 2011), *American National Standard for Electric Lamps: Specifications for the Chromaticity of Solid State Lighting Products*, [37] which established D_{uv} as a metric to quantify the distance of a source's chromaticity from the black body locus. This metric partially filled the information gap, and also drew awareness to such outlying chromaticities, which were sometimes used to increase efficacy at a given CCT. However, typical consumers do not understand this metric.

The development of LEDs for architectural lighting also led to widespread discussion of the limitations of CRI. [38] When LED products first entered the market, there was widespread concern that CRI would not work for LEDs. While some assertions were unfounded, limitations of the metric—which apply to all sources—became prominent and substantial efforts have been made to update, replace, or augment CRI. More specifically, given the color sample set and methodology used in calculating CRI, it does not always accurately portray the color rendering capability of a source, especially when there are sharp changes in the spectral content of a source over a narrow range of wavelengths. Importantly, CRI is only intended to measure fidelity, comparing a test source to a reference source; it does not indicate preference for a given SPD. This characteristic is misunderstood by many people.

Although several new color rendition metrics have been proposed that address some or all of the identified issues with CRI, to date none has been adopted by any standards or professional organization (e.g., the International Commission on Illumination [CIE] or the IES). Although the reasons for the impasse are varied, some may include the complexity of the issue, including a desire to shift to a metric that accounts for both fidelity and preference, competing opinions of separate research groups, and the potential changes to the performance ratings of existing lamps.

A separate challenge related to color quality and LEDs is effectively communicating product information to unknowledgeable consumers. It is important to include CRI, CCT, and a color scale on products to help the consumer understand the variability of LED products and increase the chance that the consumer will select an acceptable product. Currently, the Federal Trade Commission only requires the listing of CCT; however, market factors and competition may encourage the inclusion of CRI on packaging. While providing CRI, CCT, and a color scale gives consumers additional information, it still may not be enough to completely quantify color quality. Even with complete color quality data on product packaging, successfully purchasing an appropriate product relies on the education of the purchaser.

Written descriptors of color appearance are also prominent (e.g., warm white, soft white, bright white, cool white, daylight white), but their use varies among manufacturers, who sometimes refer to different CCTs with the same descriptors. Further, some terms are ambiguous, which is likely to confuse purchasers more.

A final challenge related to the wide variation of LED product color quality—and in a sense related to the ease of engineering the output of LED products—is determining what level of color quality will be acceptable to most users, resulting in mass adoption of the technology. Due to the typical performance of fluorescent lamps, a minimum CRI of 80 has become the de facto standard for commercial interiors. However, incandescent lamps with a CRI near 100 remain the benchmark for most residences, at least for now. As LEDs transition from a niche, energy-saving product to widely deployed product in a range of applications, it will be important to ensure the widespread availability of products with a desirable CRI. However, there is little research to support choosing a specific value at this time—even if ignoring the limitations of CRI as it relates to color preference. As documented in the LED Lighting Facts database, [39] most LED products intended for interior use have a CRI in the 80s, which meets ENERGY STAR and DLC QPL requirements. However, the California Energy Commission recently published the Voluntary California Quality LED Lamp Specification, which establishes a minimum CRI of 90 [40]; the widespread effect of this action remains to be determined.

3.4.4. Implications for the Future

1) Retailers should consider continuing their efforts to provide clear information regarding color quality, especially the difference between CCTs.
2) Efforts to improve the consistency of information on product packaging should continue.
3) The industry should consider establishing effective color communication tools to simplify product selection and thus improve purchaser satisfaction.
4) The lighting research community should consider establishing performance criteria for color rendering, perhaps application-dependent, that will ensure acceptability of the technology—based on color quality—to a majority of users.
5) Standards organizations should consider establishing tighter tolerances for chromaticity bins to reduce product-to-product variability at the same CCT, making it easier to specify various LED products in the same space.
6) Revised, new, or additional color rendering metrics are needed to accurately characterize the color rendering capability of all light sources, in the areas of fidelity, preference, and/or discrimination.

Ideally, such metrics would include a consumer-oriented aspect that simplifies complex information. This effort will require cooperation throughout the lighting industry, including DOE, researchers, manufacturers, and specifiers. In the meantime, CRI may be used, but its limitations must be better communicated and understood by lighting consumers.

7) The lighting research community should consider investigating thresholds for color quality acceptance, which will enable standards organizations and policymakers to establish meaningful criteria that can balance color quality with energy-efficiency goals.

3.5. Lesson 5: The Color Delivered by Some LEDs Shifts over Time, Enough to Negatively Impact Adoption in Some Applications

LEDs hold the promise of extended lifetimes, but lighting quality must be maintained over that period for the potential to be realized. Some LED-based products have exhibited unacceptable color shift after only a fraction of their rated lifetime, [41] constituting a product failure for applications where color quality is important. A lack of standard procedures for predicting and reporting color stability performance has contributed to user uncertainty—potentially limiting adoption and making it more challenging for manufacturers to provide warranty coverage for color shift.

3.5.1. Significance

Poor color maintenance can be a substantial problem in applications where color quality is important, including

- museum and gallery lighting
- architectural façade lighting
- retail display lighting
- healthcare lighting
- hospitality applications
- cove and wall wash lighting
- downlighting in commercial and residential applications.

If the color of LED products changes too much over time, to the point where it becomes unacceptable to the user or building owner, the products must be replaced. This may result in the product failing to provide the calculated payback, and will always hurt the reputation of LEDs. While many users remain unaware of the potential for color shift due to a lack of data, early installations with growing hours-of-use may soon help to identify the scope of the issue. If the issue is not addressed, LEDs may be limited to applications where color quality is less of a concern, as has occurred with other source types known to shift color over time, such as metal halide.

3.5.2. Background

Typical phosphor-coated white LEDs, the most prominent type on the market today, create white light by mixing the emission from a blue "pump" LED and a broad-emitting phosphor, with peak emission in the yellow region. If the ratio of the two emissions changes, color shift will occur along a blue-yellow axis. As LED technology has developed, the way phosphors are used has changed substantially. This has resulted in variable color maintenance performance—although it is generally improving. [42]

Physical changes—such as delamination, settling, or curling of the phosphor—are the primary cause of color shift at the LED package level. (Non-phosphor-coated and hybrid packages may have additional characteristics to consider.) The changes may occur naturally over time, but they can also be exacerbated by manufacturing defects or poor thermal management in the luminaire. Aside from physical changes to the LED package, color shift may also result from changes to other components of the complete LED product, most notably the optical system. Like lumen maintenance, heat is generally detrimental to color stability.

Early adoption of LEDs mostly took place in applications where color stability is not critical, such as roadway and parking lot lighting. In these early applications, the incumbent systems used HID sources that either offer poor color quality—such as for HPS lamps—or for which variability in color consistency and color stability was a known characteristic—such as with metal halide lamps. As LED manufacturers look to expand market penetration across a wide breadth of applications, including those with high standards for color quality, ensuring or providing warranty coverage for acceptable color stability will help to ease the concerns of purchasers.

The ability of LED products to maintain chromaticity over very long lives has been demonstrated by the L Prize competition winner, [43] but not all available products perform at that level. In fact, performance can vary significantly.

3.5.3. Challenges

While the industry reacted quickly to adopt color consistency metrics for LEDs (i.e., initial lamp-tolamp color variation) similar to those used for fluorescent sources, test procedures and relevant performance metrics for color stability have not yet been established. IES LM-80-08 details procedures for measuring of chromaticity for LED packages over time, but there is no method analogous to IES TM21-11—which is for predicting future lumen maintenance from measured data—for predicting color maintenance over time. This has resulted in a dearth of data on color maintenance, especially since color shift may begin after the LM-80-08 measurement period has concluded, leaving specifiers and consumers little option but to overlook the concern or plan for a shorter LED usable life.

The lack of prediction methodology is particularly relevant given the rapid evolution of LEDs. Measuring chromaticity to the end-of-life for LED packages is impractical given the longevity of the products and the rate at which the LED package evolves. In other words, an LED product generation is likely to be discontinued well before data collection can be completed. This challenge may abate in the future as development slows, but that time is likely far into the future.

Another challenge to the widespread availability of color shift data is the dependence of performance on the product into which an LED package is assembled. Thermal management is critical to chromaticity (and lumen) maintenance, and poorly designed lamps or luminaires can compromise the performance of an otherwise acceptable LED package. Given the wide variety of products that use a given LED package, the variation can be large and testing all combinations of products is unrealistic.

The design and manufacturing of LED packages is improving, and some data suggest that newer LED packages have significantly better color stability than older generation packages. [44] This is a step in the right direction, but without standardized and readily available data and instruction on how to use that data, specifiers may still have difficulty identifying which products are likely to perform better.

Warranty coverage could alleviate the concerns of users and specifiers, but color shift is rarely covered for currently available products, in part because of the lack of standards, but also because it is difficult to articulate how color shift should be documented. Most purchasers of LED products do not have the budget or sophisticated equipment to take photometric measurements of the products before and after a problem may occur. Further, there is no established threshold for an acceptable level of shift— although this could happen at the individual product level—and there is no consensus on the baseline for a color shift failure. That is, it must be established whether comparisons should be made to a lamp or luminaire (or its listed performance) in the new state, or to the performance of other lamps/luminaires with the same hours-of-use.

3.5.4. Implications for the Future

1) Color stability measurement and prediction methods are needed to enable performance comparisons between products.
2) There is a need for energy-efficiency programs and product information qualification standards that include color stability metrics.
3) Efforts toward establishing new methods and metrics for color stability assessment with industry and research organizations should continue. With standards and metrics, color shift can rightfully be included in a yet-to-be-developed comprehensive lifetime rating for LED products.
4) LED products need to comply with accepted tolerances for color stability.
5) Manufacturers and related industry committees should continue to share test methods and data for use by standards organizations.
6) Manufacturers should consider including color shift in product warranties.
7) Standards organizations should consider incorporating the new knowledge being generated by research organizations and industry into establishing new standards for color stability. The widespread adoption of the ANSI standards for color consistency provides a good case study for the effectiveness that a standard rating system can have.

3.6. Lesson 6: Some LEDs Flicker Noticeably, Which May Negatively Impact Adoption in Some Applications

The market adoption of LEDs is exposing the need to measure and report the flicker performance of specific products, especially since there is significant variation and severity in flicker for LED sources. Flicker is primarily a function of the LED driver design. Flicker is present in many products at full output, and dimming can induce or increase flicker, due to either the method used to implement dimming in the LED source or compatibility issues with a specific dimmer. The lighting industry needs to develop, adopt, and apply standards to limit flicker that may lead to health concerns and reduced task performance.

3.6.1. Significance

The effects of flicker range from mild to severe and affect health (headaches, eyestrain, neurological problems—including epileptic seizure [45]), diminish visual performance (reducing performance on visual tasks), cause distraction (phantom array effects seen while driving could draw the eye away from a more important situation), create work hazards (strobe effect stopping or slowing apparent motion of machinery), and provoke disruptive behaviors in individuals with autism. Due to this wide scope, flicker matters greatly for general lighting. The issue of flicker applies to almost all market segments, except outdoor applications where light levels are very low. Example areas where health or task performance are especially critical include general lighting and task lighting in healthcare facilities, classrooms, offices, retail spaces, and industrial spaces. It is less important for applications such as general illumination of roadways and parking lots. [46,47]

3.6.2. Background

Flicker is the repetitive change or modulation in luminous flux and intensity that occurs in all conventional light sources to some degree, whether perceived or not. Visible flicker is luminous modulation that is sensed and *perceived,* as opposed to invisible flicker, which is sensed but not perceived. Photometric flicker, in contrast to electrical flicker, is characteristic of the light source itself rather than being caused by disturbances in electrical input.

Conventional lighting technologies exhibit flicker in a fairly similar manner such that in many cases the issue can be remediated with appropriately designed ballasts. For example, magnetically ballasted fluorescent lamps flicker in a consistent way, and high-frequency electronic ballasts virtually eliminate that flicker. There is significant variation and severity in flicker for LED sources. The LED products most likely to flicker include AC LEDs, DC LEDs with simple/inexpensive drivers, and integral lamp LEDs on some electronic transformers. Dimming an LED source can increase or induce flicker, especially when dimmed with a phase-cut dimmer or when the driver uses a pulse-width modulation (PWM) technique to reduce LED light output.

IES has defined two metrics for flicker: Percent Flicker and Flicker Index. Percent Flicker is on a 0- 100% scale, is older and easier to calculate, and thus is more established and used. Flicker Index is on a 0-1.0 scale. Both account for average and peak-to-peak amplitudes, but only Flicker Index accounts for shape and duty cycle. Neither incorporates the critical element of frequency. For conventional sources, the maximum observed Percent Flicker is on the order of 40% and the maximum observed Flicker Index is roughly 0.15. Some LED products may have equal or better performance compared to conventional sources, while others clearly exhibit much more dramatic flicker characteristics. There are no well-defined thresholds that identify problematic flicker for specific applications or populations. [48,49]

3.6.3. Challenges

There are no standardized flicker measurement procedures in the industry, so photometric labs and manufacturers are unable to report flicker waveforms and flicker metrics consistently. Because flicker metrics are not routinely reported, lighting specifiers are forced to discover or detect flicker in other ways. Keeping in mind that the effects of flicker depend on the ambient light conditions, sensitivity of the individuals using the space, and how much eye movement is involved, other methods of discovering or detecting flicker include the following:

- Quantification:
    - ask driver manufacturers to report modulation frequency, based on instrument readings, and
    - ask manufacturers to report the dimming technique used by the LED driver. If PWM is used, and it produces 100% light modulation, then frequencies higher than 500 Hz, and possibly higher than 2000 Hz, are required to completely eliminate neurological detection. A combination of PWM and constant current reduction (CCR) dimming techniques in drivers can reduce flicker detection considerably, even at frequencies as low as 120 Hz.
- Detection:
    - observe the product in person with the same driver/transformer and dimming setting of the final installation and visually evaluate the system with flicker-sensitive clients, or
    - use a flicker wheel/spinning top under the light source; if it flickers, a checkerboard pattern will appear, or
    - rapidly wave fingers or a shiny metal rod in the light produced by the LED. Flicker will produce a stroboscopic effect, making multiple fingers or rods visible, with blank space in between the brighter images.

Developing a predictive and technology-neutral flicker metric (which in turn needs to be calibrated by human factors and medical research) and a risk assessment for difference applications will be challenging. IES and CIE are considering the development of a measurement standard and metric, and an IEEE PAR1789 committee intends to provide recommended practices for applying flicker information.

LED drivers have to strike a complicated balance between low flicker and many other factors (e.g., cost, size, efficiency, power factor, lifetime). Additional testing may be necessary to identify whether or not the interaction between LED products and some control equipment (phase-cut dimmers, in particular) results in flicker. [50,51]

3.6.4. Implications for the Future

1) Standards organizations should consider developing a measurement procedure for flicker, and a flicker metric that accounts for frequency, so manufacturers can communicate product performance.
2) Manufacturers should consider evaluating and communicating the flicker performance of their products both at full output and when dimmed, accounting for dependencies on the selection of control equipment, if applicable.
3) The lighting research community should consider performing research to establish thresholds for detection of (and perhaps objection to) flicker, risk of neurological impacts, and degradation in task performance for different applications.
4) The lighting research community should consider working together with standards organizations to develop recommended practices for specifying LED product flicker performance for different populations and applications.

3.7. Lesson 7: LEDs Can Cause Glare, Which May Negatively Impact Adoption in Some Applications

The market adoption of LEDs has grown rapidly in product categories such as streetlighting, parking lot luminaires, and directional lamps including PAR and MR lamps. [52] These lighting applications require high luminous intensities from relatively small sources. Adoption may be slowed in these and other exterior and interior applications if specifiers and users feel that LEDs produce too much glare relative to incumbent light sources that are larger and more diffuse than LEDs.

3.7.1. Significance

Adoption into important applications, many of which represent significant national energy use, may stall if LEDs earn a reputation for producing uncomfortable glare that reduces visibility. Applications where glare could limit market penetration of LEDs include

- pedestrian-scale exterior lighting
- interior high bay
- downlighting
- residential lighting
- commercial and institutional interior lighting where linear fluorescent lamps are the incumbent light source.

3.7.2. Background

The potential for glare from a light source increases both with greater lumens and with smaller source size; thus, the drive for higher light output from inherently small LEDs increases the potential for glare. Market pressure to reduce LED system costs often leads manufacturers to provide a target lumen package in a luminaire by using fewer, higher output LEDs, which also escalates potential glare. Without proper optical management, glare can be uncomfortable and even disabling for a person in a given application.

The IES luminaire classification system for exterior luminaires established a standard method for rating the backlight, uplight, and glare (BUG) from a luminaire based on standard photometric data; this rating system primarily applies to exterior roadway, area, and parking lot applications. [53] For interior applications, longstanding glare rating systems such as Visual Comfort Probability (VCP) and the Uniform Glare Rating (UGR) are generally considered not applicable or unreliable for common luminaires and systems in use today, in part because inherent assumptions in these ratings often are not accurate for contemporary systems. As a result, it is very difficult to compare the rated glare of LED systems to that of incumbent systems for applications listed above. Specifiers and users who select an LED product based only on rated data may therefore be surprised if they find that the LED system produces more glare than indicated. If this becomes a common experience, adoption of LED technology into these applications may be significantly slowed.

In exterior applications, the early adoption of LED solutions was primarily driven by energy and maintenance savings relative to HID light sources. In these applications, the incumbent light source was a high lumen, relatively small source that was mounted high above the normal line of sight. The IES BUG luminaire classification system provides a standard way for assessing glare for luminaires used in these applications. The DOE Next Generation Luminaires (NGL) outdoor competition utilized the glare ratings along with photometric and perceptual measurement from the NGL judges for evaluating LED products. [54]

3.7.3. Challenges

Considering that some of the glare issues have been escalated by the drive to increase lighting output (e.g., using higher output LEDs) or reduce cost (e.g., using fewer LEDs), one challenge to dealing with glare will be finding the balance between potentially competing goals. LED glare issues have been notable in applications where glare is problematic for incumbent technologies; thus, the challenges are not restricted to LEDs. For example, LEDs have also become a popular alternative to directional sources such as the incandescent and halogen PAR and MR lamps commonly used in retail display, museum, and other commercial and residential accent and display lighting applications. For these applications, LEDs offer substantial reductions in input power and increases in operating lifetime. The incumbent light sources produce relatively high intensities from apparent source sizes of 2-inch diameter and greater, and create glare concerns themselves. LED products designed to replace these sources typically attempt to match their intensity, but often from a much smaller apparent area, producing higher luminance and glare potential.

LED penetration into the large commercial applications using linear fluorescent lamps faces high hurdles represented by the efficacy and lifetime of the incumbent technology, and can be further hindered by increasing glare potential. Linear fluorescent lamps are relatively large, diffuse light sources, whose applications are often very sensitive to glare concerns. Even the transition from T8 to T5 fluorescent lamps created glare issues for specifiers and users, since the T5 lamp has a much smaller surface area than the T8 lamp. Even though many fluorescent applications use diffusing media in the luminaires, replacing the linear fluorescent lamps with LEDs in these luminaires can increase the potential for glare. For example, a recent DOE troffer study found that new and retrofitted troffers with odd or distracting patterns, or pronounced bright patches on lenses, were deemed more glaring than those with softer gradients. [55]

LED approaches that use mixing chambers, remote phosphors, and diffusing media in front of emitters can help mitigate the glare potential. However, because there is no widely accepted glare metric for these sources and applications, there is no convenient way for specifiers and users to assess the potential for glare from product ratings. Instead, they typically depend on mockups to evaluate glare.

These same principles apply for LED downlights, where again the LED solutions attempt to match the output of larger, more diffuse sources, creating greater glare potential. Luminaire manufacturers can use lenses, louvers, and baffles to minimize the glare, but these optical elements also reduce the optical efficiency of the luminaire, which in turn reduces the luminaire efficacy. Because luminaire efficacy is measured as part of a standard photometric test, it is widely reported on manufacturer data sheets and can be a key criterion for product selection. Conversely, the lack of a standard glare metric prevents any assessment of glare from information on a data sheet. A product with good glare control but lower efficacy can be at a competitive disadvantage as a result.

With regard to measuring glare, recent experiences have shown that the BUG ratings may not be relevant for pedestrian-scale exterior lighting systems. For example, a GATEWAY study of walkway lighting on a college campus and street and pathway lighting in a residential arts community suggests that the photometric elevation angles of 80° to 90° from nadir as glare angles in the BUG ratings are not predictive of pedestrian glare response. [56] More study is needed, but LED luminaire manufacturers may need to reexamine product light distributions, as the luminous intensity at angles from 0° to 75° may be more important for pedestrian applications.

3.7.4. Implications for the Future

1) The industry should continue work on developing LED solutions that do not increase glare relative to the incumbent technologies.
2) DOE and research organizations should consider routinely including glare assessments in product and application evaluations and demonstrations. Where a standard glare assessment methodology is not available, these organizations could work toward establishing new methods and metrics for glare assessment.
3) Manufacturers should continue to optimize their use of optical solutions that reduce source luminance, especially for applications where the incumbent light source is much lower in luminance than a typical LED.

4) Energy-efficiency programs should consider the implications of glare control when establishing efficacy standards for products. Since glare control techniques usually involve blocking light at angles most likely to cause discomfort, they almost always reduce the optical efficiency of the luminaire, which in turn reduces luminaire efficacy. A luminaire with no optical techniques for glare control will more easily meet minimum efficacy criteria than a luminaire with glare control.
5) Standards organizations should consider incorporating the new knowledge being generated by research organizations into establishing new metrics for glare assessment. The widespread adoption of the IES BUG ratings in the exterior lighting community provides a good case study for the effectiveness of a standard rating system. Similar systems are needed for the other applications discussed in this lesson.

3.8. Lesson 8: Achieving High-Quality Dimming Performance with LED Lamps Is Difficult, but Improving

Consumers have high expectations when it comes to dimming LED sources, since the technology is promoted as being inherently controllable. This is particularly true for LED lamps that are intended for applications that have historically been served by incandescent or halogen sources. While some LED sources deliver incandescent-like dimming performance, user experiences have been typically unpredictable and frequently unacceptable. This is especially true when phase-cut dimmers are used; in such instances, performance commonly depends on the specific make and model of dimmer, lamp(s), and low-voltage transformer (if applicable). Identifying combinations that will work well is complicated at best, and often impossible without a full circuit-level mockup. While new products continue to perform better than their predecessors, multiple approaches are likely necessary to fully overcome this challenge to market adoption.

3.8.1. Significance

The ability to dim LED sources holds tremendous potential for delivering even greater energy and cost savings and a wide range of other benefits, including[1]

- increased visual task performance
- enhanced ambience
- enhanced space flexibility
- lighting matched to task and/or personal preference
- demand response/load shedding
- potentially improved light source efficacy and lifetime.

The issue of dimmability is significant for both residential buildings, where phase-cut dimming is dominant, and commercial buildings, where phase-cut dimming is commonly used to control replacement lamps in various applications, such as hotel rooms, conference rooms, and retail display. [57,58]

3.8.2. Background

While there is no standard definition of "dimmable," the ability of all incandescent sources to dim smoothly and continuously down to light levels below 1% when operated by any phase-cut dimmer serves as the unofficial high-quality benchmark. The comparatively poor dimming performance of many of today's LED lamps is limiting their market adoption and damaging the perception of LED sources as inherently controllable. LED dimming performance is typically determined by the design of the LED driver. While all incandescent products are "dimmable," LED products must be designed to dim. At present, the LED lamp market contains products that are "not for use with dimmers" as well as products that make various claims of "dimmability." Such claims vary significantly; while some products advertise that they "dim like incandescent" and others specify minimum achievable light levels (e.g., "dims to 10%"), many do not further elaborate on how well they can be expected to perform.

While many LED lamps are not designed to dim as well as incandescent sources, the poor performance seen in the field is frequently not the result of LED source capability, but rather the result of LED source compatibility issues with legacy equipment. Such issues are most commonly seen when trying to dim LED lamps with phase-cut dimmers, which were fundamentally designed to control incandescent sources. LED lamps intended to replace some halogen sources (e.g., MR16) have the additional challenge of needing to be compatible with both a low-voltage transformer and a dimmer. Compatibility issues can result in a host of problems, including erratic dimming behavior, a limited ability to dim to low levels, objectionable flicker, and audible noise. [59,60]

3.8.3. Challenges

LED lamps must be designed to achieve incandescent-like dimming performance to replace those sources in many applications. While this level of performance is technically feasible, there are many barriers limiting its realization:

- Dimmable LED lamps cost more to develop and manufacture; as a result, not all products are dimmable, and those that are deliver varying levels of performance.
- Most existing dimming controls were designed for incandescent sources.
- LED sources have significantly more complex interactions with phase-cut dimmers, the most commonly deployed type of dimming control, which leads to significant technical issues.
- Variation in the installed stock of phase-cut dimmers throughout the commercial and residential markets makes addressing these interactions particularly challenging.
- Dimming approaches other than phase-control have their own challenges and market barriers, which are limiting their adoption as alternatives.
- Evidence of these challenges includes the following:
- LED lamp manufacturer claims of dimming performance are often inconsistent with observed performance. [59,60]'
- Market survey results suggest consumers are not satisfied with the observed dimming performance of many of today's products. [61]
- Laboratory evaluation of various LED lamp and phase-cut dimmer combinations has identified a wide range of technical issues that can lead to undesirable dimming performance.
- Industry has steadily acknowledged and made efforts to address the occurrence and reduce the likelihood of technical issues leading to undesirable dimming performance, and their impact on market adoption. [62,63,64]

3.8.4. Implications for the Future

1) Retailers and organizations promoting the purchase of dimmable replacements for LED integral lamps should consider increasing efforts to educate consumers and manage expectations by providing better information to buyers, alerting them to potential dimming

problems and providing clear recommendations for how to minimize the likelihood of problems.

2) Industry and organizations promoting the purchase of dimmable integral LED lamps should consider working together to develop better, clearer, and more consistent means for communicating dimming guidance to buyers. The DOE LED Lighting Facts program offers one example of both required content and format for such guidance.
3) The industry should consider working to develop LED lamps capable of high-performance dimming down to levels below 1%, with incandescent-like behavior and maintained high efficacy over the dimming range.
4) The industry should continue efforts to develop greatly improved predictability of dimming performance for specifiers and buyers.
5) Industry should continue the development of advanced dimming circuitry compatible with phase-control, focusing on improving performance with as much of the existing base of installed dimmers as possible.
6) Industry should continue efforts to develop forward-looking standards for phase-cut dimmers and lamps. NEMA SSL-7a is an important step in the right direction, but the follow-up standard (SSL7b) is still needed.
7) Industry should continue developing and promoting alternative approaches (e.g., wireless, power-line carrier), which have the potential to avoid many of the compatibility issues inherent to phase-control, and may ultimately be necessary to achieve desired performance goals.
8) DOE should continue to study, report, and challenge manufacturers to develop and market new dimming solutions.

3.9. Lesson 9: Greater Interoperability of Lighting Control Components and More Sensible Specifications of Lighting Control Systems Are Required to Maximize the Energy Savings Delivered by LED-Based Sources

The inherent controllability of LEDs could deliver unprecedented energy savings and lighting quality, optimized for the lighting application. Appropriate design is critical to maximizing the success of a lighting installation, and even more so when lighting controls are integrated into the installation. However, a well-designed lighting control system may not ultimately meet user needs and expectations if it is improperly installed and not fully commissioned, or is difficult to modify or augment to meet user needs that have changed or become more realized over time. The delivery of energy savings and lighting performance that both maximizes the potential offered by LEDs and meets user needs over time will likely require continuous improvement in the way lighting controls are designed and specified. The ability and willingness of lighting users to learn how to operate and modify sophisticated lighting controls systems varies significantly. Lighting control specifications should focus on delivering solutions that suit both the use and the user. Greater interoperability of lighting control components can increase the chances of proper installation and commissioning, and reduce user risk during specification.

3.9.1. Significance

The controllability of LEDs, including the ability to adjust output level, color temperature or chromaticity, and undergo frequent on/off switching without detriment, significantly enhances the energy savings LEDs are capable of providing. While adjustments that result in lower power levels directly impact energy savings, adjustments that result in improved lighting quality or otherwise increase user satisfaction can also impact energy savings if they increase the adoption of LED technology. The incumbent technology for many applications is not easily dimmed or requires expensive upgrades to enable dimming; in many cases, this has stymied the deployment of controls for those applications. For example, difficulties dimming HPS and HID sources have hindered the use of lighting controls in outdoor applications, and the significant cost difference between standard and dimmable fluorescent ballasts has limited the energy savings achieved from deploying controls in some environments.

No lighting technology has been capable of delivering variable color temperature or chromaticity as effectively and (potentially) efficiently as LED. LED technology is poised to bring high-performance, (potentially) low-cost control of output level, and/or color to many lighting applications for the first time. Integrating lighting controls with a communication network offers additional opportunities to provide value to users and perhaps save money and additional energy. The ability to track real-time energy use and report failures can reduce maintenance costs and save energy in some instances, such as when outdoor lighting dayburners are identified. As the LED lighting market matures, maximizing the energy savings delivered by the installation of LED lighting systems, which if designed well could have unprecedented field life, will become increasingly dependent on maximizing the successful installation of lighting controls that suit the use and the user.

3.9.2. Background

While LEDs are inherently controllable, they are not the only controllable lighting technology. Incandescent sources dim wonderfully and can be controlled by inexpensive equipment (e.g., phase-cut dimmers). However, incandescent sources are inefficient and lose efficacy when dimmed. More efficient HID technologies (e.g., HPS and metal halide) are difficult to dim and can require significant periods following a power shut-off to return to full brightness. Fluorescent sources offer high efficacy and excellent dimming performance when equipped with high-performance (and historically expensive) ballasts, but struggle to deliver the lighting quality required by some applications. Adjusting the color temperature or chromaticity of all these incumbent technologies requires the use of filters, which fundamentally reduce their efficacy.

Lighting controls have been developed and deployed to varying degrees of success for decades. Most control technologies were developed for specific lighting technologies. For example, phase-control was developed for incandescent sources and 0-10V and digital addressable lighting interfaces were developed primarily to control linear fluorescent sources. Multiple strategies exist for using lighting controls to deliver energy savings (e.g., scheduling, daylight harvesting, occupancy sensing, task tuning, and personal tuning) and combining multiple strategies allows for maximum savings. The energy savings and user satisfaction delivered by market-available lighting control solutions has varied significantly; a recent meta-analysis of real-world results in commercial buildings showed a potential variation in energy savings of roughly 40 to 50% for any given strategy, and almost no impact from implementing multiple strategies. [65]

Despite significant research efforts, predicting how much energy savings can be achieved through the implementation of one or more well-established lighting control strategies remains difficult. As summarized by the Lighting Controls Association "...achieving energy savings estimates in practice may require commissioning, including a written controls narrative, verification equipment is installed and aimed in accordance with approved documents, programming and calibration, functional testing, systems manual, end-user training and a plan for periodic recalibration." [66] In addition, factors beyond the choice of strategy and how effectively that strategy is implemented (e.g., how well the selected system works with other [often existing] equipment and how well-suited the solution is for the building and space characteristics, use, and occupants—all of which can change over time) often influence the success of a lighting control installation and the energy savings achieved.

3.9.3. Challenges

Lighting controls cover a broad spectrum of capability and sophistication, ranging from simple wall-box dimmers controlling one lighting circuit to networked building-wide systems that can interface with other (non-lighting) systems and be accessed from anywhere through the internet. Further, the ability and willingness of lighting users to learn how to operate and modify lighting controls varies significantly, which presents a significant matching challenge. Ensuring that a lighting control system well suited to its use (i.e., building, space, occupants) is specified, properly installed, and fully commissioned does not necessarily lead to persistent energy savings. Use characteristics change or sometimes simply become better understood over time, often requiring modification of or augmentation to a lighting controls system. Users are less likely to make changes that prove to be too difficult, time-consuming, or costly, which leads to reduced user satisfaction and ultimately reduced energy savings over time. In extreme cases, systems may be intentionally defeated or simply turned off.

Lighting control systems have long suffered from the assumption that they are simply "plug and play." This perception poses two major challenges. If insufficient time and attention is given to the design, installation, or commissioning processes, many lighting control systems will not meet user expectations and/or not function optimally. For example, when implementing an occupancy sensing strategy, important design considerations include the placement of the occupancy sensors, the luminaire dimmed level setting when in the "unoccupied" mode, and the occupancy sensor delay setting that determines how long the luminaire stays in the "occupied" mode following a detection. Appropriate time and attention given to the design process can add 30% or more to the energy savings achieved by an LED system installed without such controls. Conversely, failure to adequately address these basic design considerations can essentially render the installed controls and their associated investment useless. [67]

A second set of challenges posed by the "plug and play" misconception include the assumptions that a) lighting control components are compatible with existing infrastructure; b) lighting control components from different manufacturers are interoperable (i.e., work well together); and c) lighting control components that seem to perform the same function and even have similar specifications are, in fact, interchangeable. The frustration caused by these challenges can lead to significant market adoption barriers. For example, the compatibility issues that many LED integral lamps have with the installed base of phase-cut dimmers has led to many early adopters to not trust the dimmability claims of these products. At present, most lighting control systems require the use of proprietary hardware and/or software, thereby requiring the potential user to make a substantial investment in products from a single vendor that then locks them in for future purchases. The user must continue to buy from that same vendor if they want new system components to work well with those purchased previously. This lack of interoperability increases user risk when considering new installations, especially in instances where user needs are not fully understood at the time of specification or are likely to change over time. If the chosen vendor cannot support changing needs, the user may be faced with the decision to start over from scratch or live with an existing, increasingly unsuitable system.

Many lighting controls are marketed as complying with one or more "standards." However, lighting specifiers and users often do not fully understand what some of these standards ensure. For example, the 0-10V standard does not specify when, or even whether, a luminaire should turn off; as a result, when two luminaires are presented with the same control signal, one may turn off, while the other may go to a low lighting level. Similarly, the DALI standard has not historically required compliance testing, leading to different manufacturers developing different "versions" of DALI products, which are often not interoperable.

Interoperability and interchangeability concerns may be less important for relatively small, self-contained lighting systems (e.g., those servicing a conference room or a single parking structure). Conversely, larger-scale building-level or street lighting systems can include thousands of luminaires and other connected equipment spread across numerous locations with different use and geographic characteristics. The higher complexity and variation found in such systems leads to significantly higher risk for users who locks their future to a single vendor. The challenges presented by the lack of interoperability may increase over time as more systems become interconnected in support of net-zero building, smart city, smart grid, and intelligent transportation initiatives.

3.9.4. Implications for the Future

1) The lighting industry should consider continuing its efforts to develop and refine educational and certification programs aimed not only at selecting and designing for control strategies, but at ensuring that the specified lighting control equipment suits both the use and the user, and is correctly installed and fully commissioned. In some cases, simpler is better. Systems with fewer compatibility issues are more likely to be installed correctly and fully commissioned. Similarly, systems that are easier to modify or augment may lead to more persistent energy savings.
2) LED products that deliver on their promise of long lifetime will be in the field, in some cases, for the next 20 years or more. Once installed, the incremental cost required to return to the installation site to add controls may be prohibitive or not in line with payback expectations. The lighting industry should consider exploring all opportunities to maximize the number of LED product installations that incorporate sensibly specified controls. At the same time, the lighting industry should recognize that installations that incorporate controls that are not properly installed; are not fully commissioned; require a degree of user sophistication beyond user capabilities or interest to maintain, modify, or augment; or otherwise do not meet user expectations will only damage the reputation of lighting controls and limit their adoption. In cases where the installation of lighting controls is truly not immediately practical, lighting specifiers and lighting system users should strive to identify and require the installation of "control-ready" LED products that minimize the future hardware, software, and labor costs for installing a control system.
3) The specification of interoperable lighting controls can significantly reduce user risk when first installing a system and maximize the chance that the installed system will continue to meet expectations as user needs change or simply become better understood. Manufacturer consortiums should considering continuing their efforts to develop open-standard specifications and compliance testing programs that allow lighting control products to be brought to market that both offer new features and deliver some level of interoperability. Careful, ongoing attention should be given to balancing the sometimes-conflicting goals of delivering deeper levels of interoperability while allowing for manufacturer competition that encourages innovation and

the development of new features. Deeper levels of interoperability should be delivered as the market matures to drive user adoption.

4) Efforts to bring interoperability to the lighting control market are already underway within the ZigBee Alliance, LonMark International, the TALQ Consortium, the Connected Lighting Alliance, and others. Lighting specifiers, lighting control system designers, and lighting system users will likely increasingly demand interoperability (especially if they are installing LED sources) and could help manufacturer consortiums determine what level of interoperability is required at various stages of market maturity. Liaisons with user organizations (e.g., the MSSLC) that can represent the collective voice of specific user communities can aid in both the development of interoperable specifications and the education of what they do and do not deliver. Energy-efficiency organizations that focus their lighting control incentive programs on interoperable equipment may be able to develop cost-effective programs more efficiently and accelerate the development and deployment of such equipment.

3.10. Lesson 10: Lack of LED Product Serviceability and Interchangeability Has Created Market Adoption Barriers in Certain Sectors

Despite the promise of long life, there remains much uncertainty about the lifetime and reliability of LED products. As a result, designers are often uncomfortable specifying products that do not have serviceable components. Furthermore, even in cases where existing LED products offer acceptable performance and cost savings, some users hesitate to buy LED products that cannot be upgraded while the technology is still advancing so rapidly.

3.10.1. Significance

Many lighting clients have strict performance and cost expectations. Designers who fear lifetime and reliability issues and their resultant liability may delay the specification of LED products, thereby forgoing the significant energy savings already available for some applications. Similarly, clients and designers who feel compelled to wait for better performance, lower costs, and/or new features may avoid LED products (and their associated energy savings) until the market is effectively commoditized. Serviceability reduces initial risk, and interchangeability offers protection against obsolescence. This issue applies primarily to the commercial market, especially the architectural and hospitality sectors.

3.10.2. Background

Early experiences with some LED products has shown that, while the LED packages themselves may be long-lived and have acceptable lumen maintenance, driver failure or shifting color can cause what initially looked like a successful installation to fall short of client expectations. The time and cost required to address failures can be a function of luminaire construction. In some cases, the entire luminaire must be removed and replaced, which may be considered wasteful by clients concerned about sustainability, and may be untenable for expensive architectural or decorative luminaires. Alternatively, modular luminaire designs may allow for only the failed component to be removed and replaced. Such serviceable luminaires may not only reduce the risk of unmet lifetime expectations, but may also have lower life-cycle costs.

The continual evolution of LED technology presents challenges of its own for some lighting clients. The reality that tomorrow's products will perform better and perhaps have more capabilities than today's products makes it difficult for some designers to decide when is the right time to start specifying LED products, especially for clients who do not consider life-cycle costs. Furthermore, some lighting clients expect to be able to take advantage of evolving LED technology by, for example, upgrading some luminaire component to deliver higher efficacy or the ability to control color temperature in a couple of years. In both cases, luminaires designed with interchangeable components, such as LED modules or light engines, facilitate such upgrades much more than those that do not.

3.10.3. Challenges

Modular luminaire designs have three potentially significant drawbacks. The first is cost; designing any product to have serviceable or interchangeable parts requires the use of plugs, sockets, and perhaps additional wiring and housing materials, all of which increase manufacturing cost. Modularity can also potentially create performance constraints. Any design parameter that is "locked down" to facilitate modularity reduces design freedom, which potentially reduces performance.

Finally, the ability to exchange a failed component for a new one or an old component for a newer, higher performing one also presents the opportunity to install a component that is not well-suited to a given luminaire, potentially leading to reduced performance or safety hazards. While some modular interfaces are very well defined, and their use in products tightly controlled, other (typically more common and familiar) modular interfaces are used in widely varying products, without sufficient consideration for mismatches. Some users may assume that any component that fits into a given receptacle is designed for the system containing the receptacle. In others, a user may not know how to properly install or secure a component into a luminaire. Examples of what can go wrong include

- installing medium screw-based LED lamps into sealed fixtures designed for incandescent lamps, resulting in lamp temperatures that exceed design specifications
- installing double-end powered bi-pin LED T8-replacement tubes or fluorescent T8 tubes (back) into a troffer re-wired for single-end powered LED T8-replacement tubes, resulting in potential safety issues
- installing an LED module that requires a thermal interface material into a socket that requires re-application of such material, or never contained such material.

3.10.4. Implications for the Future

1) The industry should consider adopting standardized modular interfaces. At present, manufacturers are incorporating modularity into their LED luminaire designs to different degrees, and for different purposes. Some manufacturers design modules with a focus on speeding the product development time for traditional or less sophisticated luminaire designers, or see modularity as a means for serviceability (perhaps by trained personnel only), but not interchangeability. Others seek to cater to users familiar with lamp-based luminaires by developing LED modules or light engines that may be exchanged by general users, either to address failures or deliver improved performance or new features. Some manufacturers use proprietary interfaces, while some are working together (e.g., through the Zhaga Consortium) to define standard interfaces. Finally, even when modular design is common for a particular component, as is the case for LED troffers that typically contain a modular LED driver, taking advantage of said modularity may not be straightforward. There is considerable variation in driver design, as well as LED boards and arrays, and there are no industry standard methods for characterizing them.

2) Manufacturers should consider developing products with serviceable or interchangeable components. The challenges presented by the lack of LED product serviceability or interchangeability—or in some cases by the inconsistent implementation of modularity—are likely to persist for the foreseeable future, while LED technology continues to evolve rapidly, driving manufacturers to explore new design approaches and system partitioning.
3) As LED technology matures, market preferences—which may be application- or sector-specific—will likely emerge. In the meantime, applications and market sectors that are most affected by the lack of serviceability and interchangeability should be identified, and manufacturers encouraged to appropriately focus their early efforts at delivering both.

3.11. Lesson 11: Existing Lighting Infrastructure Limits the Full Potential of SSL; More Effort Is Needed to Open the Doors to New Lighting Systems and Form Factors

Almost the entire SSL market remains focused on products that fit into the existing infrastructure of legacy lighting products. This is true for lighting product form factors, customer expectations, and lighting product functionality. While this is a necessary and expected consequence of introducing a radically new technology into a mature market, the current market focus sharply limits the potential for the new technology. The sooner the market begins to transition away from its legacy focus, the sooner SSL can deliver deeper energy savings, and generally better lighting performance possible only with luminaires designed from the ground up to handle LED light sources.

3.11.1. Significance

The unique characteristics of LEDs, including their small size, ease of controllability, directionality of emissions, and durability, among many others, offers unprecedented flexibility for how SSL product designers can deliver lighting and related services to lit spaces. Novel configurations and form factors have high potential for

- deeper energy savings
- better lighting quality

- lower lighting costs
- longer life
- more comfortable and attractively lit spaces
- more light where it is needed or wanted, and less light where it is not
- better controllability with regard to dimming, frequent on/off switching, spectral tuning, beam shape, and other characteristics.

This issue applies across all lighting markets, but especially to those where consumers or non-lighting professionals are purchasing lighting products. It also applies in markets where lighting infrastructure is rarely upgraded or replaced. This means the problem is most prevalent in the residential and small commercial markets, as well as in other commercial markets with high sensitivity to first costs and reluctance to try unfamiliar product designs. [68]

3.11.2. Background

Perhaps nothing better illustrates this issue than the current heavy market focus on integral LED lamps intended as replacements for incandescent light bulbs (A lamps). Because incandescent A lamps were inexpensive, offered good quality lighting, and were quick and easy to replace, they evolved into a standard industry product. They had become the standard light source for a large variety of luminaires, from table lamps, to ceiling mounted fixtures, porch lights, downlights, and desk lamps. A huge infrastructure of lighting fixtures, along with customer knowledge and expectations, was built up over the century of incandescent A lamp prevalence. That infrastructure developed around an omnidirectional light source that can withstand high-temperature operation, has a standardized electrical interface (Edison base), has a very low purchase price, functions as a simple resistor on electric circuits, and is nearly identical from product to product except for light output—which varies according to a very easy-tounderstand metric (lamp power).

It is therefore only natural that manufacturers developed LED products to fit into this infrastructure and paradigm. But to make LED products function well in that infrastructure, product designers have been forced to sacrifice product life, lumens, controllability, cost, energy efficiency, and lighting quality. As a simple example, consider an integral LED replacement for an incandescent bulb in a table lamp. Side-emitting lumens are wasted because that same lamp might also be installed by a buyer in a ceiling fixture requiring high side emissions. The Edison base largely prevents use of the fixture as a heat sink, causing the lamp to run hotter than it otherwise would. If that bulb is used in a dimmable fixture, sophisticated electronics are needed to make it dim reasonably well on the existing wall dimmer, increasing the cost of the bulb. Energy is wasted because lumens are wasted, and because the LEDs have to be driven relatively hard to produce high light output from the limited bulb size. Finally, the product life is shortened due to high-temperature operation caused by dense packaging of LEDs and electronics within that small form factor. [69]

As with the A lamp example, legacy lighting infrastructure and customer expectations greatly limit product design freedom and functionality across the lighting landscape, from fluorescent troffers to halogen MR16 accent lights.

3.11.3. Challenges

Manufacturers will make only what customers are willing to buy, and customers cannot easily give up their old habits, expectations, and investment in existing hardware. This is true for most mature product markets into which radically new technologies are introduced; for example, early cars had to function (and look) a lot like horse buggies. Upgrading electrical infrastructure in a home or business costs money, requiring the services of qualified electricians, and significant renovation work to walls and ceilings to access wiring. Electrical systems have to comply with the National Electrical Code and local building codes. Even if building owners wanted to renovate and prepare for use of SSL immediately, it is not clear what would be required or advisable, because SSL technology, products, control systems, and market availability all continue to change and evolve rapidly. Over time, newly constructed buildings will incorporate electrical infrastructure that is better suited to full application of digital SSL technology (along with a range of other communications, controls, and grid response features). Decades of infrastructure transition are to be expected.

3.11.4. Implications for the Future

1) Government organizations, codes and standards bodies, and specifiers need to be mindful that their lighting requirements may (albeit unintentionally) restrict product form factors, functionality, and system operation. Careful development of these requirements opens the door to innovation and better product designs.
2) Energy-efficiency program operators should consider explicitly allowing for non-conventional form factors and functionality, and to the extent possible, move away from program designs built around the concept of one-for-one product substitutions.

3) Product buyers are more likely to buy products not compromised by the legacy infrastructure if those products offer compelling functionality. For example, lamps controlled wirelessly circumvent the limitations of having to communicate a dimming signal over the wires that conduct power to lamps, but also allow those lamps to be controlled for a wide variety of other characteristics, such as on/off schedules, color point, and light beam shapes.
4) Manufacturers, lighting educators, DOE, and others could induce earlier acceptance by customers of lighting products by frequently raising the issue of what is possible with SSL when unconstrained by existing infrastructure. Manufacturers with innovative product designs at the ready will be in the best position to leverage this new opportunity.

3.12. Lesson 12: Programs that Provide Ways to Identify Quality LED Products Have Helped Support Market Adoption

Individual consumers, utility program managers, retailers, facility managers, and lighting designers all need help sorting through and understanding LED product performance data, but few have the time and expertise to evaluate every product. Several programs have emerged to help: CALiPER, LED Lighting Facts, ENERGY STAR, DLC, and others verify product performance information and make at least some of that information publicly available. These programs have helped support market adoption of LED products by assuring stakeholders that products have been tested and reviewed by an independent entity. These programs have simultaneously discouraged the practice of inflating performance claims, as was rampant in the early days of LED product development.

3.12.1. Significance

The LED lighting market is varied and complicated, with new products and applications appearing continuously. LED lighting cannot be categorized as one single light source type that will replace another discrete light source. LEDs are better characterized as the next phase of lighting, which could potentially replace all other electric light source types, although not yet all lamp wattages or all lighting applications.

Because of this product diversity, LED performance cannot be generalized. It depends on the lighting application, the product category, and individual product design. Qualified product lists and objective third-party LED product information are relevant for all lighting sectors, including residential, commercial, municipal, institutional, and industrial facilities. Program support is essential to help utilities, energy-efficiency programs, retailers, and consumers evaluate LEDs. Verified test data and independent product qualification is needed to

- help utilities determine which products are eligible for incentives and other promotion
- help designers screen products for consideration
- help retailers determine which products to stock
- help facility lighting buyers determine which products to purchase
- inform consumers of basic performance characteristics
- encourage manufacturers to align performance claims with tested, verified results.

3.12.2. Background

When LED products were first introduced in the market, many had unrealistic and unsupported performance claims. Until the IES photometric test procedure, LM-79, was published, there was no standard method for testing LED products, creating much confusion about which metrics should be reported. [70,71,72] Some luminaire manufacturers reported the claimed efficacy or output of the LEDs used in their products, a value based on instantaneous measurements taken during LED package production. Equivalency claims (with respect to the type and output of conventional lamps for which they were claimed to be suitable substitutes) were often highly exaggerated, with products claiming to replace a 40- watt or 60-watt incandescent lamp producing less than one-quarter of the light output of those benchmarks. In the first two rounds of CALiPER testing in 2007, less than 15% of the products had light output claims within 30% of tested values. [71] By the third round of CALiPER testing, however, the proportion of products with reasonably accurate light output claims (within 10% of tested values) had risen to 30%, suggesting that the publication of these discrepancies between output claims and test results had an immediate effect on the market. [72] Of products CALiPER tested in 2012, about 67% had output claims within 10% of the tested value. [73] Once LM-79 testing became available from independent test laboratories, the industry had a standard method to measure LED product performance. This enabled the development of minimum voluntary performance criteria for qualification by ENERGY STAR and the

DLC, and the LED Lighting Facts database of verified product performance. As of 2013, more than 2,800 LED integral lamps and 2,100 LED luminaires are ENERGY STAR-qualified, more than 31,000 products are DLC qualified, and more than 10,000 products are listed on LED Lighting Facts.

3.12.3. Challenges

LED product offerings continue to increase very rapidly. Keeping pace with advancements in LED technology and product design is a challenge for qualification programs that have minimum performance requirements. Keeping information current is a challenge for all programs, due to the frequent updating and redesign of products based on new LED component availability, pricing, competitive demands, and other factors. The frequent need to update information presents time and cost burdens to participating manufacturers and to third-party programs themselves. [74,75,76]

3.12.4. Implications for the Future

1) Efficiency program managers should consider prioritizing the development of custom options that help users identify quality LED products. These programs should target energy-efficiency performance that will produce significant energy savings if the technology is widely adopted in the market.
2) Efficiency program requirements need to keep pace with LED technology development, balancing energy efficiency, lighting quality, and cost considerations.
3) Testing and documentation requirements should be streamlined as much as possible, while still maintaining confidence in qualified product performance claims.
4) Wherever possible, data should be shared across the various performance verification programs to save time and cost for all participants.
5) It is not clear how long such programs will be necessary, as the LED lighting market continues to evolve and mature. Continued rapid change in the technology, product offerings, and integrated systems indicate objective third-party information will continue to be needed for at least the next 5 years.

CONCLUSION

The effective application of the lessons learned from past experiences with technology and product development can significantly improve the outcomes of similar future endeavors when these lessons are effectively applied. DOE and others have clearly applied many of the lessons learned during the development and commercialization of CFLs to the development and market introduction of energy-efficient LED lighting systems for general illumination, as evidenced by proactive, collaborative efforts to establish performance testing, standards, and education programs. Partly as a result of applying these and many other past lessons learned, the development and market introduction of LEDs has gone much more smoothly than CFL market introduction (see Figure 4.1). However, the unique technological characteristics of LED lighting have presented a host of new challenges and lessons.

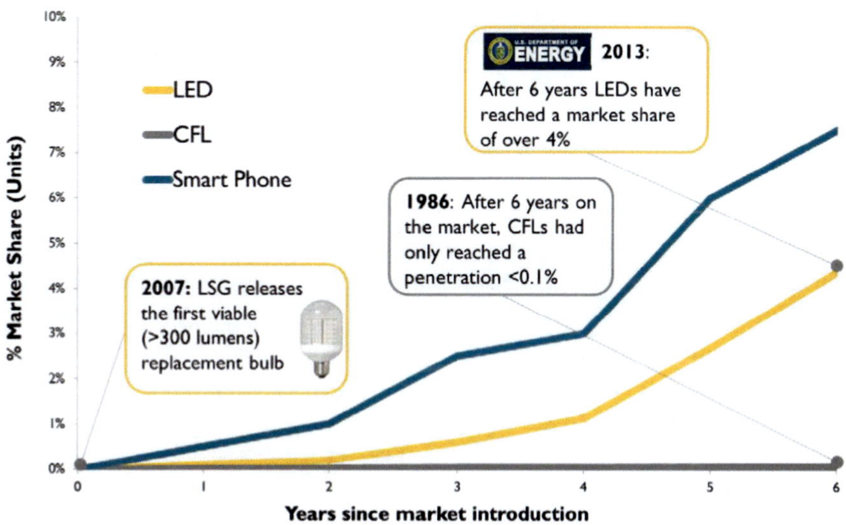

Figure 4.1. Comparison of Market Share Increases after Product Introductions for CFLs, LEDs (Lamps and Luminaires), and Smart Phones. Source: Navigant Consulting, Inc.

This report documents early challenges and lessons learned in the SSL market and summarizes early actions taken by DOE and others to avoid potential problems anticipated based on lessons learned from the market introduction of CFLs. LED technology and product performance cannot be easily generalized. The technology is capable of very high luminous efficacy, color rendering, power quality, beam control, dimming performance, lifetime, and other attributes. The actual performance of any given product depends on the target application, pricing and operating cost structure, form factor, maintenance requirements, and other constraints and tradeoffs. A continuing challenge for the lighting community is the difficult tradeoffs among various performance attributes and price.

Based on early LED market experience, this study identifies and characterizes 12 key lessons distilled from DOE's SSL Program results with a focus on areas for which ongoing challenges exist and/or useful information can be applied going forward. These lessons correspond with technological challenges related to performance and lifetime of LEDs, color quality and measurement, flicker, glare, dimming and control, serviceability; challenges related to developing LED product lines and families; and complications and limitations experienced when trying to fit LEDs into existing lighting infrastructure. Lessons from the many programs that have helped support market adoption are also summarized.

SSL is revolutionizing lighting products, systems, and practices. One can envision a future in which the lights in buildings, homes, and outdoor areas use just a fraction of the electricity now used for lighting; have excellent quality, durability, and ability to adjust the amount and color of light; are integrated with information and communications systems; and are responsive to the needs of the electrical grid. Realization of the full benefits of SSL—in terms of lighting quality, energy efficiency, environmental sustainability, controllability, and value-added technology integration—is a tantalizing possibility, but by no means assured. This early assessment of the LED general illumination market is intended to aid in the continuous course corrections needed to reach that full technology potential.

End Notes

[1] "Solid-state lighting" (SSL) refers to the type of lighting that uses illumination sources that include semiconductor light emitting diodes and encompasses several technologies including inorganic light-emitting diodes (LEDs) and organic light-emitting diodes (OLEDs). This report addresses market experience to date with the introduction of lighting products using LEDs. Market issues related to OLED lighting products are outside the scope of this report.

[2] Sandahl LJ, T Gilbride, M Ledbetter, HE Steward, and C Calwell. 2006. *Compact Fluorescent Lighting in America: Lessons Learned on the Way to Market.* Prepared by Pacific Northwest National Laboratory for the Building Technologies Program, Office of Energy Efficiency and Renewable Energy, U.S. Department of Energy, Washington, D.C.

Available at http://apps1.eere.energy.gov/buildings/publications/pdfs/ssl/ cfl_lessons_learned_web.pdf.

[3] This report addresses market experience to date with the introduction of lighting products using LEDs. To avoid making the report overly complicated, the term "LED" is sometimes used generically to describe LED technologies in general and the terms SSL and LED are sometimes used interchangeably. Specific discussion of LED subassemblies and systems (LED drivers, engines, etc.) are only referenced when the discussion directly relates to these subassemblies.

[4] DOE. 2013a. *Solid-State Lighting Research and Development: Multi-Year Program Plan.* Prepared for Lighting Research and Development, Building Technologies Office, Office of Energy Efficiency and Renewable Energy, Washington, DC.

[5] Sandahl, et al., *Compact Fluorescent Lighting in America: Lessons Learned on the Way to Market.*

[6] DOE. 2008. *Multi-Year Program Plan for FY'09-FY'14, Solid-State Lighting Research and Development.* Prepared for Lighting Research and Development, Building Technologies Program, Office of Energy Efficiency and Renewable Energy, U.S. Department of Energy, Washington, DC.

[7] DOE and IALD. 2008. Memorandum of Understanding Between the United States Department of Energy and the International Association of Lighting Designers. U.S. Department of Energy, Washington, DC and International Association of Lighting Designers, Chicago, IL. Available at http://apps1.eere.energy.gov/buildings/publications/pdfs/ssl/iald-doe_mou-final.pdf.

[8] *ENERGY STAR Program Requirements for Solid State Lighting Luminaires*, version 1.0, took effect September 30, 2008. *ENERGY STAR Program Requirements for Integral LED Lamps*, version 1.0, took effect August 31, 2010.

[9] EPA. 2012. *ENERGY STAR Summary of Lighting Programs: September 2012 Update*. Prepared by ICF International, Washington, DC. September 2012. Available at http://www.energy star.gov/ia/partners/downloads/ 2012_ENERGY_STAR_Summary_of_Lighting_Programs. pdf.

[10] CALMT. 2013. *Statewide Lighting Market Transformation Program Report.* June 2013. Annual Report. California Utilities Lighting Market Transformation Stakeholders. Available at http://www.lightingmarkettransformation.com/wp-content/uploads/2013/06 /2013-Statewide-Lighting-MarketingTransformation-Program-Report.pdf.

[11] Where "MR" refers to a multifaceted reflector and "PAR" refers to a parabolic aluminized reflector.

[12] DOE's GATEWAY demonstrations evaluate and showcase real world applications of high-performance LED products for general illumination in a variety of commercial and residential applications.

[13] DOE. 2009. *CALiPER Summary of Results: Round 9 of Product Testing*. October 2009. Office of Energy Efficiency and Renewable Energy, U.S. Department of Energy, Washington, DC. Available at http://apps1.eere.energy.gov/buildings/publications/pdfs/ssl/caliper_round-9_summary.pdf.

[14] MSSLC. 2013. *Model Specification for Adaptive Control and Remote Monitoring of LED Roadway Luminaires, V1.0.* Municipal Solid-State Street Lighting Consortium, Office of Energy Efficiency and Renewable Energy, U.S. Department of Energy, Washington, DC. Available at http://www1.eere.energy.gov/buildings/ssl/controlspecification.html.

[15] DOE. 2007a. *CALiPER Summary of Results: Round 1 of Product Testing*. March 2007. Office of Energy Efficiency and Renewable Energy, U.S. Department of Energy, Washington, DC.

Available at http://apps1.eere.energy.gov/buildings/publications/pdfs/ssl/cptp_round_1_testing_results_summary.pdf.

[16] DOE. 2007b. *CALiPER Summary of Results: Round 2 of Product Testing*. August 2007. Office of Energy Efficiency and Renewable Energy, U.S. Department of Energy, Washington, DC. Available at http://apps1.eere.energy.gov/buildings/publications/pdfs/ssl/cptp_round_2_summary_final_draft_8-15-2007.pdf.

[17] DOE. 2012a. *CALiPER Year in Review 2012*. December 2012. Office of Energy Efficiency and Renewable Energy, U.S. Department of Energy, Washington, DC. Available at http://apps1.eere.energy.gov/buildings/publications/pdfs/ssl/caliper_2012-review.pdf.

[18] DOE's LED Lighting Facts® program showcases LED products for general illumination from manufacturers who commit to testing products and reporting performance results according to industry standards. The program provides information essential to evaluating SSL products available on its website, http://www1.eere.energy.gov/buildings /ssl/ledlighting facts.html.

[19] DOE. 2012b. *Life-Cycle Assessment of Energy and Environmental Impacts of LED Lighting Products, Part I: Review of the Life-Cycle Energy Consumption of Incandescent, Compact Fluorescent, and LED Lamps*. February 2012, updated August 2012. Office of Energy Efficiency and Renewable Energy, U.S. Department of Energy, Washington, DC. Available at http://apps1.eere.energy.gov/buildings/publications/pdfs/ssl/2012_LED_Lifecycle_Report.pdf.

[20] DOE. 2012c. *Life-Cycle Assessment of Energy and Environmental Impacts of LED Lighting Products, Part 2: LED Manufacturing and Performance*. June 2012. Office of Energy Efficiency and Renewable Energy, U.S. Department of Energy, Washington, DC. Available at http://apps1.eere.energy.gov/buildings/publications/pdfs/ssl/2012_led_lca-pt2.pdf.

[21] DOE. 2013b. *Life-Cycle Assessment of Energy and Environmental Impacts of LED Lighting Products, Part 3: LED Environmental Testing*. March 2013. Solid-State Lighting Program, Building Technologies Program, Office of Energy Efficiency and Renewable Energy, U.S. Department of Energy, Washington, DC. Available at http://apps1.eere.energy.gov /buildings/publications/pdfs/ssl/2013_led_lca-pt3.pdf.

[22] Current market data provided in Dan Chwastyk's (Navigant Consulting) presentation, "DOE's Market Introduction Workshop." November 13, 2013. Available at http://apps1.eere.energy.gov/buildings/publications/pdfs/ssl/chwastyk_mktevolution_portland2013.pdf.

[23] IES LM-79-08. *Approved Method: Electrical and Photometric Measurements of Solid-State Lighting Products*. Illuminating Engineering Society, New York, New York.

[24] DOE, CALiPER Summary Reports, 2006-2008. http://www1.eere.energy.gov /buildings/ssl/report-archives.html.

[25] DOE. 2012d. *Solid-State Lighting Manufacturing R&D Workshop Report*. Office of Energy Efficiency and Renewable Energy, U.S. Department of Energy, Washington, D.C. Available at http://apps1.eere.energy.gov/buildings/publications/pdfs/ssl/sanjose2012_report.pdf.

[26] Next-Generation Lighting Industry Alliance and DOE. 2011. *LED Luminaire Lifetime: Recommendations for Testing and Reporting, Second Edition*. June 2011. Available at http://apps1.eere.energy.gov/buildings/publications/pdfs/ssl/led_luminaire-lifetime-guide_june2011.pdf.

[27] DOE. 2013c. *Solid-State Lighting Technology Fact Sheet: Lifetime and Reliability*. Office of Energy Efficiency and Renewable Energy, U.S. Department of Energy, Washington, DC. Available at http://apps1.eere.energy.gov/buildings/publications/pdfs/ssl/life-reliability_fact-sheet.pdf.

[28] DOE. 2013d. *CALiPER Application Summary Report 20: LED PAR 38 Lamps*. November 2012, Addendum September 2013. Prepared by Pacific Northwest National Laboratory for the Office of Energy Efficiency and Renewable Energy, U.S. Department of Energy, Washington, DC. Available at http://apps1.eere.energy.gov/buildings/publications/pdfs /ssl/caliper_20_summary.pdf.

[29] IES LM-80-08. *Approved Method: Measuring Lumen Maintenance of LED Light Sources*. Illuminating Engineering Society, New York, New York.

[30] IES TM-21. *Projecting Long Term Lumen Maintenance of LED Light Sources*. Illuminating Engineering Society, New York, New York.

[31] DOE. 2013e. *Hammer Testing Findings for Solid-State Lighting Luminaires*. Office of Energy Efficiency and Renewable Energy, U.S. Department of Energy, Washington, DC. Available at http://apps1.eere.energy.gov/buildings/publications/pdfs/ssl/hammer-testing_Dec 2013. pdf.

[32] DOE. 2012e. *LED Color Characteristics*. Prepared by Pacific Northwest National Laboratory for the U.S. Department of Energy, Washington, DC. Available at http://apps1.eere.energy.gov/buildings/publications/pdfs/ssl/led-color-characteristics-factsheet.pdf.

[33] Sandahl, et al., *Compact Fluorescent Lighting in America: Lessons Learned on the Way to Market.*

[34] CIE 15-2004. *Colorimetry*, 3rd edition. International Commission on Illumination, Vienna, Austria.

[35] CIE 13.3-1995. *Method of Measuring and Specifying Colour Rendering of Light Sources*. International Commission on Illumination, Vienna, Austria.

[36] DOE, 2012e. *LED Color Characteristics.*

[37] ANSI ANSLG C78.377-2011. *Specifications for the Chromaticity of Solid State Lighting Products*. National Electrical Manufacturers Association, Rosslyn, VA.

[38] Houser KW, M Wei, A David, MR Krames, and XS Shen. 2013. "Review of measures for light-source color rendition and considerations for a two-measure system for characterizing color rendition." *Optics Express* 21(8).

[39] Lighting Facts database available online at: http://www.lightingfacts.com/.

[40] CEC. 2012. *Voluntary California Quality Light-Emitting Diode (Led) Lamp Specification: A Voluntary Minimum Specification for "California Quality" LED Lamps*. California Energy Commission, Sacramento, CA. Available at http://www.energy.ca.gov/2012publications /CEC-400-2012-016/CEC-400-2012-016-SF.pdf.

[41] DOE. 2013f. *Color Maintenance of LEDs in Laboratory and Field Applications*. Prepared by Pacific Northwest National Laboratory (PNNL-22759) for DOE SSL Program's Gateway Demonstrations, Buildings Technology Office, Office of Energy Efficiency and Renewable Energy, U.S. Department of Energy, Washington, D.C. September 2013. Available at http://apps1.eere.energy.gov/buildings/publications/pdfs/ssl/2013_gateway_color-maintenance.pdf.

[42] DOE. 2013f. *Color Maintenance of LEDs in Laboratory and Field Applications.*

[43] DOE 2013f. *Color Maintenance of LEDs in Laboratory and Field Applications.*

[44] DOE 2013f. *Color Maintenance of LEDs in Laboratory and Field Applications.*

[45] The range of flicker that can affect seizures is 3 to 70 Hz. This is not the normal range of flicker for LED products in the U.S.; however, some past products that used unidirectional strings of LEDs with AC produced 60 Hz flicker, and some holiday lights still use this approach. In addition, the failure mode of some LED systems can produce low frequency flicker.

[46] DOE. 2013g. *Flicker*. Building Technologies Office, Office of Energy Efficiency and Renewable Energy, U.S. Department of Energy, Washington, DC. Available at http://apps1.eere.energy.gov/buildings/publications/pdfs/ssl/flicker_fact-sheet.pdf.

[47] Poplawski M and N Miller. 2011. *Exploring flicker in Solid-State Lighting: What you might find, and how to deal with it*. Prepared by Pacific Northwest National Laboratory for the U.S. Department of Energy, Washington, DC. Available at http://www.e3tnw.org /Documents/2011%20IES%20flicker%20paper%20poplawski-miller-FINAL.pdf.

[48] DOE, 2013g. *Flicker*.

[49] Poplawski and Miller. *Exploring flicker in Solid-State Lighting: What you might find, and how to deal with it*.

[50] DOE, 2013g. *Flicker*.

[51] Poplawski and Miller. *Exploring flicker in Solid-State Lighting: What you might find, and how to deal with it*.

[52] DOE. 2013h. *Adoption of Light-Emitting Diodes in Common Lighting Applications*. April 2013 (Revised May 2013). Prepared by Navigant Consulting, Inc. for Solid-State Lighting Program, Building Technologies Office, Office of Energy Efficiency and Renewable Energy, U.S. Department of Energy, Washington, DC. Available at http://apps1.eere.energy.gov/buildings/publications/pdfs/ssl/led-adoption-report_2013.pdf.

[53] IES. 2011. *The Lighting Handbook*, 10th edition. IES-HB-10-11, Illuminating Engineering Society, New York, NY.

[54] NGL provides information on its evaluation criteria and judging process online by year and category. See http://www.ngldc.org/.

[55] DOE. 2013i. *CALiPER Exploratory Study: Recessed Troffer Lighting*. . Prepared by Pacific Northwest National Laboratory (PNNL-22348) for the Office of Energy Efficiency and Renewable Energy, U.S. Department of Energy, Washington, DC. March 2013. Available at http://apps1.eere.energy.gov/buildings/publications/pdfs/ssl/caliper_recessed-troffer_2013.pdf.

[56] DOE. 2013j. *Pedestrian Friendly Outdoor Lighting*. Prepared by Pacific Northwest National Laboratory (PNNL23085) for DOE SSL Program's Gateway Demonstrations, Buildings Technology Office, Office of Energy Efficiency and Renewable Energy, U.S. Department of Energy, Washington, DC. December 2013. Available at http://apps1.eere.energy.gov /buildings/publications/pdfs/ssl/2013_gateway_pedestrian.pdf.

[57] DOE. 2012f. *LED Dimming: What you need to know*. Office of Energy Efficiency and Renewable Energy, U.S. Department of Energy, Washington, DC. Available at http://apps1.eere.energy.gov/buildings/publications/pdfs/ssl/dimming_webcast_12-10-2012.pdf.

[58] MSSLC, *Model Specification for Adaptive Control and Remote Monitoring of LED Roadway Luminaires*.

[59] DOE. 2012f. *LED Dimming: What you need to know*.

[60] DOE. 2013k. *Dimming LEDs with Phase-Cut Dimmers: The Specifier's Process for Maximizing Success*. Prepared by Pacific Northwest National Laboratory (PNNL-22945) for DOE SSL Program's Gateway Demonstrations, Buildings Technology Office, Office of Energy Efficiency and Renewable Energy, U.S. Department of Energy, Washington, DC. October 2013. Available at http://apps1.eere.energy.gov/buildings/publications /pdfs/ssl/2013_gateway_dimming.pdf.

[61] McGaraghan MD et al. 2013. "LED Lamp Quality: Codes and Standards Enhancement (CASE) Initiative for PY 2013: Title 20 standards Development." *Analysis of Standards Proposals for LED Replacement Lamp Quality*. Docket #12-AAER-2B. July 29, 2013.

[62] NEMA LSD-49. *Solid State Lighting for Incandescent Replacement—Best Practices for Dimming*. National Electrical Manufacturers Association, Arlington, VA. Available at http://www.nema.org/Standards/Pages/SolidState-Lighting-for-Incandescent-Replacement-Best-Practices-for-Dimming.aspx.

[63] NEMA SSL 6. *Solid State Lighting for Incandescent Replacement—Dimming*. National Electrical Manufacturers Association, Arlington, VA. Available at http://www.nema.org/Standards/Pages/Solid-State-Lighting-forIncandescent-Replacement-Dimming.aspx.

[64] NEMA SSL 7a. *Phase Cut Dimming for Solid State Lighting: Basic Compatibility*. National Electrical Manufacturers Association, Arlington, VA. Available at http://www.nema.org/Standards/Pages/Phase-CutDimming-for-Solid-State-Lighting-Basic-Compatibility.aspx.

[65] Williams AB et al. 2012. "Lighting Controls in Commercial Buildings." *Leukos* 8(3):161-180. Available online at: http://www.ies.org/leukos/samples/1_Jan12.pdf.

[66] Dilouie C. 2013. "Estimating Energy Savings with Lighting Controls." From Lighting Controls Association online publication, available at http://lightingcontrolsassociation.org/estimating-energy-savings-with-lightingcontrols.

[67] DOE. 2012g. *Use of Occupancy Sensors in LED Parking Lot and Garage Applications: Early Experiences*. Prepared by Pacific Northwest National Laboratory (PNNL-21923) for DOE SSL Program's Gateway Demonstrations, Buildings Technology Office, Office of Energy Efficiency and Renewable Energy, U.S. Department of Energy, Washington, DC. October 2012. U.S. Department of Energy, Washington, D.C. Available at http://apps1.eere.energy.gov/buildings/publications/pdfs/ssl/2012_gateway_sensors.pdf.

[68] These conclusions are based on a series of discussions and input gathered during SSL past workshops and conferences. More information on past conferences is available on the SSL Program website: https://www1.eere.energy.gov/buildings/ssl/past_conferences.html.

[69] Based on input gathered during SSL past workshops and conferences. See SSL Program website for more information on past workshops: https://www1.eere.energy.gov/buildings/ssl/past_conferences.html.

[70] DOE, 2007a. *CALiPER Summary of Results: Round 1 of Product Testing*.

[71] DOE, 2007b. *CALiPER Summary of Results: Round 2 of Product Testing*.

[72] DOE. 2007c. *CALiPER Summary of Results: Round 3 of Product Testing*. October 2007. Office of Energy Efficiency and Renewable Energy, U.S. Department of Energy, Washington, DC. Available at http://apps1.eere.energy.gov/buildings/publications/pdfs/ssl/caliper_round_3_summary_fnl.pdf.

[73] DOE. 2012a. *CALiPER Year in Review 2012*.

[74] EPA. 2013. *Energy Star Certified Light Bulbs and Light Fixtures*. U.S. Environmental Protection Agency, Washington, DC. Available at www.energystar.gov.

[75] DLC. 2013. *Design Lights Consortium*. Design Lights Consortium, Northeast Energy Efficiency Partnerships, Inc., Lexington, MA. Available at http://www.designlights.org/.

[76] DOE. 2013l. *LED Lighting Facts*. Office of Energy Efficiency and Renewable Energy, U.S. Department of Energy, Washington, DC. Available at http://www.lightingfacts.com/.

In: Solid-State and Compact Fluorescent Lighting ISBN: 978-1-63482-881-9
Editor: Joan L. Thompson

Chapter 2

COMPACT FLUORESCENT LIGHTING IN AMERICA: LESSONS LEARNED ON THE WAY TO MARKET*

L. J. Sandahl, T. L. Gilbride, M. R. Ledbetter, H. E. Steward and C. Calwell

EXECUTIVE SUMMARY

Compact fluorescent lamps (CFLs) were introduced in the 1970s; however, it has taken more than 30 years for them to gain widespread recognition in the U.S. residential lighting market. Government and utilities have invested significant public and ratepayer funds in promoting the technology because CFLs represent an effective energy efficienct resource: they decrease lighting energy consumption by one-half or more for each incandescent lamp replaced. While CFLs still have only about 2% of the national market in terms of unit sales, they have achieved higher market penetration in some parts of the country, including California (5%) [43] and the Pacific Northwest (12%) [44].

This report reviews efforts to increase market acceptance of CFLs and barriers to that acceptance. The report was prepared by the Pacific Northwest

* This is an edited, reformatted and augmented version of a June 2006 report, PNNL-15730, prepared under contract by Pacific Northwest National Laboratory for the U.S. Department of Energy.

National Laboratory for the U.S. Department of Energy's Office of Building Technologies, Emerging Technologies Program. Lessons to be learned from this study of CFLs are identified in hopes of assisting future market introduction efforts for other promising energy- efficient lighting technologies.

Lessons learned from reviewing the efforts of the many parties involved in increasing consumer acceptance of residential CFL lighting and from studies of CFL manufacturers, retailers, and consumers are summarized below. These lessons are grouped into three categories: those pertaining to technology, to marketing, and to program design. A fourth category lists lessons learned from recent interviews PNNL conducted with CFL manufacturers Philips, GE, Osram Sylvania, and TCP regarding how their experiences would apply to the market introduction of LEDs (light emitting diodes) or other new lighting technologies.

Technology

- Introduce new lighting technology first in niche applications where benefits are clearly defined, and consistent with buyer needs.
- Be aggressive about dealing with technology failures that affect main benefit claims.
- Performance is more important than appearance (e.g. twister CFLs).
- Know and admit technology limitations.
- Manufacturers and energy-efficiency groups should coordinate to establish minimum performance requirements.

Marketing and Consumer Education

- Identify and focus first on niche markets where the benefits of the new technology make sense.
- Focus message where the technology can meet/exceed expectations.
- In marketing literature show comparisons of savings to the standard technology, e.g., 1 CFL = 10 incandescents.
- Back up long-life claims with guarantees.
- Let consumers see new technology - use lit in-store displays and see-through packaging.

- Work toward consistent, industry-wide terminology. Identify and avoid terms with negative connotations.
- Be specific about benefits. How long do they last? How much do they save?
- Shift consumer focus from product price to product value. Build a message of value (costs more/worth more).
- Use advertising messages that celebrate distinct advantages of new technology and disadvantages of old.
- Conduct in-store product demonstrations.
- Promote products through several different mediums to reinforce familiarity.
- Target training programs/awareness campaigns to market channels such as builders, designers, and retailers.
- Provide retailers with many education tools including brochures, posters, demonstrations, and wall displays. Support retailer efforts with utility bill stuffers and product demonstrations at local fairs, home shows, and energy shows.
- Use public service advertising.
- Offer consumers a pathway for answering questions through telephone hotlines and websites printed on product packaging.
- Arrange shelf displays by lighting application rather than manufacturer and identify good/better/best options that correlate to longer product life and greater energy efficiency.
- Strive to make new lighting technologies available through market channels where products are typically purchased (e.g., lighting purchases are often made at grocery stores).
- Avoid the short-term fix of a rebate unless it is tied to an overall campaign that includes an education campaign.
- Use mass media, both paid (TV, radio, newspaper, and magazine ads) and unpaid (host press events, tie news releases to current events).

Program Design

- Study market structure to see how best to introduce a new technology.
- Don't rely on utility giveaway programs that bypass normal distribution routes or undermine retail sales.

- Avoid give-aways and programs that obscure retail price, leading to sticker shock when consumers return for repeat purchase.
- Incentives are typically most effective when directed at manufacturers versus consumers.
- Tie retailer incentives to retailer training.
- Focus product marketing on attribute that is most important to consumers (e.g., energy savings and long life for CFLs).
- Require action on part of consumer, in give-away programs, for higher installation and retention rates, and greater consumer awareness, e.g., make customers mail in a request card to get the free bulb, don't just hand them out door-to-door to customers who may not want them.
- Use utility or manufacturer field representatives as link between manufacturers and retailers; they can educate and train retailers, set up displays, distribute promotions, and iron out problems.
- Encourage and applaud enthusiastic and ongoing retailer participation.
- Delay program launch rather than introduce inferior products; first impressions are long lasting.
- Design multi-year programs around the lighting season (Sept to April) not the calendar year.
- Invest in visible, attractive point-of-purchase displays and signage.
- Educate retailers and provide them with tools - in-store displays, signage, brochures, more informative packaging, etc. (See 2.5.1, 4.2)
- Coordinate incentive programs; avoid overlapping, inconsistent promotions.
- Join forces with others in national energy-efficiency programs (e.g., ENERGY STAR).

Application to LED and Other New Lighting Technologies

- A lot of consumer research is needed to determine what the consumer does and does not know before the initial product launch so that the launch is done right the first time.
- Accurate incandescent equivalency on packaging is critical.
- Rely as much as possible on retailers for customer education. Product packaging can also be a very powerful way to convey product benefits.

- Industry collaboration, perhaps through NEMA, would be helpful, though difficult to achieve given the large number of manufacturers.
- Coordinate with energy-efficiency programs once products are available but don't start before products are ready.
- Don't rely on giveaways and coupons or other programs that confuse consumers about the actual retail price.
- Avoid market introduction programs that distribute products outside normal retail channels, for example utility mail-order programs.
- Performance claims must be accurate. Don't launch a product until performance issues are ironed out.
- Initial education and performance issues will be more difficult to iron out if many manufacturers are involved in the initial introduction of LEDs.
- Pricing is critical but tricky - low enough to encourage consumer demand, high enough to generate profit for the retailer and manufacturer.
- Education, of both consumers and retailers, is critical.
- Understand that many people will not try a new product until price drops to a range near that of existing products providing similar functionality.
- Niche marketing is the best approach for now.

1.0. INTRODUCTION

Compact fluorescent lamps (CFLs) were introduced in the 1970s; however, it has taken more than 30 years for them to gain widespread recognition in the U.S. residential lighting market. This report, prepared by Pacific Northwest National Laboratory (PNNL) for the U.S. Department of Energy's Appliances and Emerging Technologies Program, reviews efforts to increase market acceptance of CFLs and barriers to that acceptance. Drivers and effective marketing messages are discussed. Lessons to be learned from this study of CFLs are identified in hopes of assisting future market introduction efforts for other promising energy-efficient technologies, such as LEDs.

1.1. Background

The first commercially available fluorescent lamps were introduced in the late 1930s [2] and compact fluorescent lamps were introduced to the U.S. market in the late 1970s (see Appendix A for a brief history of this development). Fluorescent lamps convert more of the power used to usable light; consequently, they are three to six times more efficient than incandescent lamps [2]. But compact fluorescents didn't gain widespread acceptance or even acknowledgment among U.S. consumers until after 2001, despite efforts by electric utilities and CFL manufacturers in several parts of the country.

1.1.1. Early Utility Programs - 1980s-1990s

Beginning in the late 1980s, incentive regulation allowed many utilities for the first time to generate revenue from investments in energy efficiency programs, [24] and utilities eager to undertake new demand side management opportunities latched onto the fledgling CFL technology [8].

The pursuit of residential lighting programs made sense. Studies showed that residential lighting comprised a significant percentage of U.S. energy expenditures. A 1997 study for the California Energy Commission noted that, in California, residential lighting energy use was about two-thirds the size of commercial lighting energy use and 8% of the overall statewide electricity use [20]. According to the National Association of Home Builders, the average U.S. home size increased 22% from 1987 to 2001 [40] and a 1999 study noted that average installed watts per home had increased by an average of 100 watts per decade in California, attributable to that steady increase in the size of homes [20]. A 2002 report for DOE estimated average household lighting energy use at 1,946 kWh/yr [39], double a 1993 estimate by the Energy Information Administration, which put average annual lighting energy consumption at 940 kWh/yr per household [40, although other researchers have noted that this may be a low estimate]. A later California study also noted an increase in the average number of lamps (bulbs) per home from 34 in 2000 to 41 per home in 2005 [37].

Throughout the 1990s, CFLs were promoted aggressively in certain markets of the country through a variety of individual utility programs (see examples of utility programs in Section 3). Utilities promoted CFLs as a replacement for the standard A line incandescent lamp, doubtless wishfully eyeing the savings potential from whole-house changeovers from incandescents to CFLs. Unfortunately the early-model CFLs were not quite ready for prime time, with a host of technical challenges including bulkiness, low light output, and inconsistent performance, still to be worked out before they would be considered comparable to incandescents in any application.

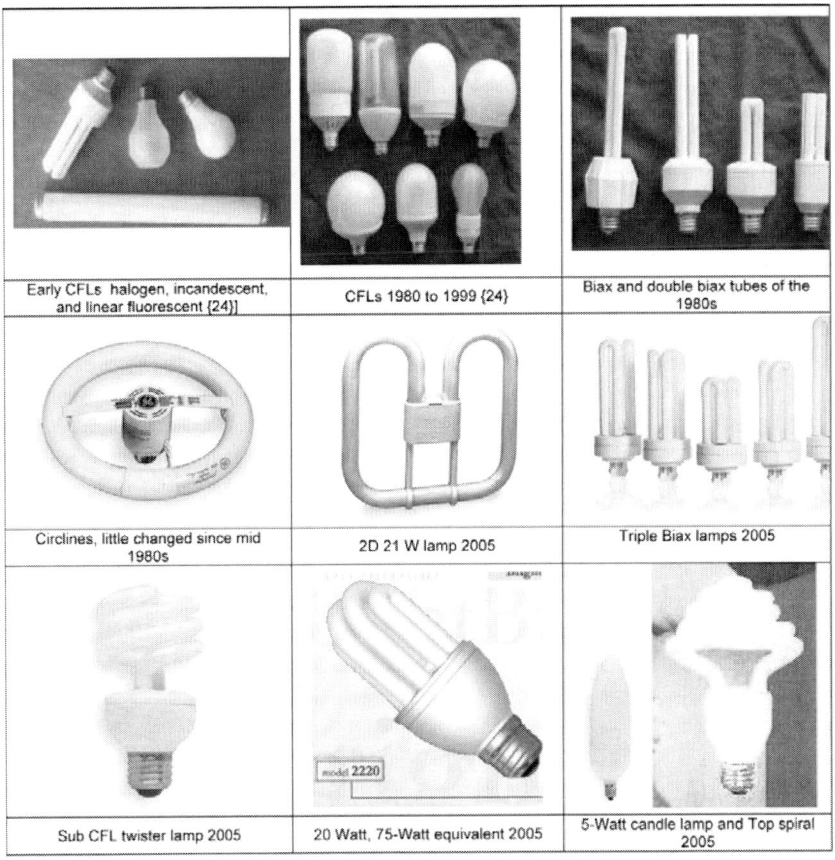

Figure 1.1. Evolution of Compact Fluorescent Lamp Designs. CFLs have come in a variety of shapes and sizes, but the overall trend has been downsizing from the early 2-feet-long U-tubes of the 1970s (which were basically a 4-ft long, 1.5-inch-diameter T-12 bent in half) to the T6 0.5-in. diameter U-tube (6-in. lamp length plus ballast) of the mid 1980s, to the triple tube or triple biax lamps (4.9 in. lamp, 6 in. with ballast) and double U tube (3-in. lamp length plus ballast) designs of the mid 1990s, to the 4.0-in. twister lamps that first appeared in 1995.

1.1.2. Technological Improvements

Many of the technical barriers were resolved over time with technological advances. The greenish tint of early fluorescents, due to the halophosphate phosphors used, was eliminated with the development of rare earth phosphors in the late 1970s. These phosphors provided warmer colors, better color rendition, increased light output, and higher efficacy, which enabled reductions in lamp length [6]. The ballasts, which fluorescent lamps need to provide starting voltage and regulate current during lamp operations, were also in need of improvement. Early magnetic ballasts were replaced in the late 1980s with electronic ballasts that enabled reduced weight, lower energy consumption, less humming, and less flicker than their magnetic predecessors [21]. Overall size continued to decrease as engineering improvements were made in lamp and ballast designs (see Figure 1.1). Unfortunately not all of these improvements were made in every model all at once, especially in lower cost lamps.

1.1.3. Slow U.S. Market Adoption

Even with resolution of these technical barriers, U.S. residential adoption of CFLs seemed slow in coming, especially compared to European and Asian markets. A two-year study concluded in 1996 found that 56% of households in the Netherlands, 50% of households in Germany, 46% of households in Denmark, and 20% of households in the United Kingdom had CFLs installed compared to under 10% in the United States [23]. In Japan by 1993, 80% of household lighting was provided by fluorescent lamps (including tubes and circulars) while in the U.S. CFLs accounted for less than 1% of incandescent sales. [10]

As Calwell stated in a 1999 report to NRDC, "The reasons for the difference between the western European experience and the U.S experience are due to many factors, including national coordination of promotional efforts, different cultural attitudes about resource consumption and, to some extent, higher electricity prices. One particularly interesting example is from Poland, which has electricity rates similar to the US, but with much lower average household income. CFL penetration in Poland went from almost nothing in 1993 to over 30% nationally by 1998. The reasons appear to be a combination of broad manufacturer advertising and a national, market based, CFL promotion program. Despite national differences, there are valuable lessons from programs in other countries that can be incorporated into U.S. program designs." [23]

What was holding the U.S. market back? In addition to the technical barriers, there were other consumer barriers, some persisting from the 1980s through to today. These included high CFL price compared to low incandescent prices and relatively low electricity prices, lack of awareness of and knowledge about CFLs, consumer buying habits, negative perceptions and skepticism about fluorescent lighting, and factors particular to the US lighting market]Other market forces and market players had roles in both hindering and helping to increase acceptance of CFLs. These barriers, drivers to acceptance, and lessons learned are described in the sections below.

1.2. Overview of this Report

Consumer, manufacturer, retailer, and market barriers are described in Section 2 of this report. Drivers for increased acceptance of CFLs including utility programs, regional programs, nationwide efforts like ENERGY STAR, the impact of the Western electric energy crisis of 2001, and the impact of California's Title 24 are described in Section 3. Section 4 describes key marketing messages and trends in advertising, promotion, and consumer education and lessons learned from these activities.

Section 5 gives a snapshot of where the market is now in terms of consumer awareness, consumer satisfaction, availability, and market share. Section 6 describes future challenges, what barriers remain, what areas are being pursued, and new California Title 24 requirements. Section 7 describes implications of the CFL experience for LED lighting. Section 8 lists lessons learned. Section 9 is references. Appendix A provides a brief history of CFL development. Appendix B is a timeline of important events in the evolution of CFLs in the United States.

2.0. Barriers and Issues

Many barriers and issues have been identified that in one way or another impeded the adoption of CFLs in the U.S. residential market. It should be noted that some of these issues are inherent in the adoption of most new technologies.

2.1. Early Consumer Barriers – Technical Challenges and Performance Issues

Throughout the 1980s, and even into the early '90s, consumers who tried the early models of CFLs found plenty not to like about the new technology. And early utility programs may have unwittingly added to consumer dissatisfaction by distributing lower cost (and lower performing) models to keep program costs down. [24].

In addition to the high initial cost of CFLs, there were a number of other barriers to increased use of CFLs encountered by consumers in the 1980s and '90s, including:

Size/fit – In 1990 CFLs were still so big and bulky consumers had to replace lamp shades, change lamp harps, or get modification kits to make the lamps fit in some of their fixtures. A 1993 study showed that even with modifications the CFLs still wouldn't fit in over 60% of the fixtures in an average home [7, 9,13].

Performance – Several utility surveys and consumer focus group studies pointed to performance issues such as the humming, buzzing, and flickering associated with early magnetic ballasts, delayed start, lack of dimmability, and poor outdoor performance as reasons for consumer dissatisfaction [8,17,20] NEEP specifically stopped promoting more compact and high performance electronically ballasted units in favor of magnetically ballasted units in the early 1990s, because its engineers imposed stringent power factor and THD requirements that the early electronically ballasted models could not meet.

Light quality, output, brightness – Poor light levels and color rendition were an early and ongoing source of dissatisfaction and a primary reason for early removal of installed CFLs. [7,9,16,20] 50% of those who installed then removed CFLs in a 1990 Pacific Gas and Electric program felt that the 27-watt CFLs they purchased had not lived up to the program's claims that they were equivalent to a 100-watt incandescent. [7] While some aspects of light quality might be due to consumer perceptions, light quality studies by Rensselaer's Lighting Research Center confirmed that color measurements vary between and even within a manufacturers' CFL product lines. [36]

Early failure – early bulb burnouts were a major source of dissatisfaction for participants in utility programs, especially since long life was one of the advertised strengths of CFLs [17,24]. This would be problematic for any technology, but especially for one that demanded such a price premium compared to the conventional product it replaced.

2.2. Consumer Barriers

In addition to the technical barriers identified above, several other barriers that could be seen as directly related to the consumer have been identified. These include price, awareness and knowledge of CFLs, attitudes and fears about CFLs (and fluorescent lighting in general), skepticism of product claims, and buying habits. The fact that so many of these barriers relate to consumers' perceptions about the product should be a wake-up call to the important role that marketing and consumer education can play in increasing consumer acceptance of a new technology.

2.2.1. Price

Price was a huge initial barrier with retail prices of $25 to $35 per bulb common in the mid 1980s [24]. In study after study consumers pointed to price as their number one obstacle to purchasing a CFL [8,9,10,11,13]. Even if the products lasted as long as 10 to 15 incandescent bulbs, they cost more than buying that many incandescents and required the purchase to be made all at once, rather than over time. To put it bluntly, how do you convince someone to mail order a bulky, odd looking "fluorescent device" they've never seen in operation for $30 when they can buy a four pack of familiar incandescents for a dollar at the neighborhood grocery store?

Many utility studies indicated that $10 was a magic price point. As an EPRI consumer focus group reported, consumers considered prices over $10 to be "outrageous." [8] But a 1993 Southern California Edison study found 70% of respondents wouldn't pay more than $5 for a CFL [11] and a 1992 San Diego Gas and Electric study found many unwilling to pay more than $2 for a CFL. Utility give-away programs and rebates helped many customers clear the price barrier to at least try a CFL but in some ways these programs were also detrimental to consumer opinions. According to a 1991 EPRI focus group study, the deep discounts and free CFLs offered by utilities only served to confirm many customers' feelings that CFLs "were drastically over-priced" [8]. Consumers who did purchase early CFLs at retail prices usually purchased lower cost (and often inferior) products [8].

Due in part to various utility promotions, CFL prices varied widely. A Southern California Edison study found that the same model of CFL carried five different prices ranging from $6.97 to $19.99 at five different stores in their service territory on the same day in January 1997 [18]. This wide variation in price made it difficult for consumers to put a value on the new product. A 1999 report written by Ecos Consulting for NRDC noted that when asked about the cost of incandescent bulbs, most Southern California Edison customers correctly guessed prices of about $2 for a pack of four. Likewise, most correctly guessed that the cost of a typical four-foot fluorescent tube would be $3 to $5. However, when the researchers questioned those already familiar with CFLs about the typical cost of a CFL, more than half responded that they did not know and they wouldn't even guess at a price [24].

As is often the case with new products, prices did eventually come down from an average of $19 in 1996 [17] to $11 in 1999 [23] to $6 or less in 2003 [35], thanks in part to increasing competition, increased demand and production spurred by utility and national marketing programs, the West Coast energy crisis, and other drivers discussed in Section 3.

2.2.2. Awareness/Knowledge

When Madison Electric started its CFL program in 1990, only 5% of its customer base had tried CFLs and only 4 of the 100 retailers it polled carried them [11]. A review of studies done in 1992 and 1993 with three different utilities showed 40% of their customers were unaware of CFLs and close to 80% had never purchased one. [13] This review concluded that "consumer awareness and misperception about CFL performance, along with lack of retail availability, may be bigger obstacles to CFL adoption than physical limitations". [13] A 1998 study of over 500 Southern California Edison customers found only 15% had CFLs installed [18]. A 1999 study of 500 customers in the Pacific Northwest, where CFLs had been promoted by local utilities for years, found only 57% had heard of CFLs and of those only 42% had tried them. [17, 22]

Clearly, there was a great deal of early confusion about CFLs. In an EPRI 1991 consumer focus group study, 1 in 3 consumers who initially defined themselves as non-CFL users discovered in the course of the discussion that they had actually tried compact fluorescent lighting but were confused about the correct terminology. [8] In an SDG&E 1992 focus group some participants were unsure whether CFLs could be installed upside down. [9] Many were also unaware that CFLs could be used in typical incandescent fixtures. [9] Adding to the confusion was the lack of a consistent name for the new product. The consulting firm Heschong Mahone Group noted in 1999 [20] that manufacturers referred to their CFL products by very different terminology. These have included CFBs[13], triple tubes, biax bulbs, triple CFLs, triple biax lights, quad tubes, Earth Light SLs [12], cf bulbs [9], SL-lamps, and PL-lamps. [6]

Several studies showed that consumers were unsure about the terms "watts" versus "lumens;" consumers consistently underestimated the amount of energy CFLs save or how long they last. [8,9,18,23,24] A 1995 survey by Phillips of 1,000 consumers found 42% did not know the difference between incandescent and fluorescent bulbs. [23] Even as late as May 2006, lumens versus watts was still a confusing issue for consumers, with consumers wondering if they could replace incandescents with higher wattage equivalent CFLs, for example using a 100 watt CFL equivalent in a 60 watt fixture (personal comm.. My Ton, Ecos Consulting May 2, 2006.)

2.2.3. Attitudes, Fears, and Perceptions

CFLs had an image problem that was hard to overcome. In many consumer focus groups, the very word "fluorescent" invoked connotations such as "harsh, cold, glaring, flickering, buzzing, artificial, and ugly" and fluorescent lighting was associated with eye strain, noise, greenish skin tones, and institutional settings [8,9,20]. Consumers expressed concerns about the safety of CFL lighting in general and its potential impacts on their health. Some participants in a focus group thought the CFLs they were given to try looked unsafe and that the tubes should have some kind of covering. [9] Surveys have shown that some customers would refuse to install CFLs under any circumstances due to health and safety concerns. Some individuals maintained that fluorescent lighting causes headaches and even cancer, regardless of the ballast and phosphor technologies employed.

Several early consumer studies showed consumers were skeptical of packaging claims of long life and energy savings. Consumers could not relate to the claims of 6,000 or 8,000 hour bulb lives [9]. Consumer who had tried CFLs had difficulty believing the energy savings claims because most households were only willing to purchase one or two test lamps due to their high initial cost; thus, the savings would not make enough of a dent in the monthly bill to be noticeable [9]. Several consumers specifically questioned why long life claims were not backed by a guarantee [8,9]. In a 1998 telephone survey of over 200 Pacific Northwest customers who had not heard of CFLs, when asked if they would purchase an energy-efficient bulb that costs $10 but lasts 10 times as long as incandescent bulbs and was good for the environment, one-fourth said they would not because they were skeptical about environmental claims and didn't think light bulbs could have anything to do with helping the environment [22]. Focus group studies by EPRI in five major U.S. cities in 1991 found that some consumers distrusted utility motives and actually felt that reduced energy usage would result in higher rates. [8] This sentiment was echoed by participants in a 1992 SDG&E focus group study. [9] Marketing aimed at overcoming these attitudes would be necessary to win over significant portions of the market.

An early San Diego Gas and Electric study (1992) proved that actual experience with CFLs can change consumers' perceptions. SDG&E gave a group of customers six very different looking styles of CFLs ranging from circline to biax, to tubes covered with a plastic dome to resemble an oversized A-lamp. At first introduction, before using the bulbs, several consumers said that manufacturers should make all CFLs look more like incandescents. After trying the bulbs for 4 weeks, the customers independently chose a new favorite, the circline model, this time basing their choice not on appearance but on light quality. The 30-watt circline gave a light output equivalent to nearly 150 watts, the highest of any of the bulbs in the program. [9]. And, as later LBNL research would show, it directed that light primarily upward and downward, where it would be of greatest use to those relying on table and floor lamps to provide enough light for reading. This posed a conundrum for early program implementers, since a circline is arguably not a new compact fluorescent at all, but simply a conventional, large-diameter, linear fluorescent lamp curved into a circular shape and self-ballasted. The new, high-tech, electronically ballasted, small CFLs were being given large utility rebates and extensive promotion, but were gaining less consumer support than the much cheaper, technically "inferior" circular fluorescents already on the market.

2.2.4. Purchasing Behavior

Americans typically purchase light bulbs at grocery stores; however, until recently it was not likely that CFLs could be found there in any significant numbers or at competitive prices [8,11,17]. This lack of availability of CFLs at the location where consumers are making light bulb purchases doubtless hurt early sales. Strong promotion of CFLs through mass merchandise and home improvement stores has changed some consumer bulb buying habits. But old habits die hard, especially when buying incandescents is as convenient as buying a gallon of milk. Habit is what made 26% of those surveyed return to buying incandescents after they had purchased, and been satisfied with CFLs, according to a 2004 report prepared by ECONorthwest for the Northwest Energy Efficiency Alliance. [35]

2.3. Market Availability Barriers

This consumer practice of buying light bulbs at grocery stores collided with another barrier that could be termed both a retailer barrier or a manufacturing barrier or even a market structure barrier. It has to do with the way incandescent light bulbs have traditionally been sold in this country.

According to a November-December 1994 *Home Energy* article, "As a nation we buy most of our light bulbs at the grocery store. The incandescent bulb industry in the United States has been dominated by three manufacturers: General Electric, Philips, and Osram-Sylvania. Grocery stores typically carry only one of these three brands because of exclusivity agreements, so when customers go to the store they only have one brand to choose from."[14] One study asserted that this lack of competition limits the potential for an innovative product to challenge traditional buying patterns. [15] Other smaller manufacturers that wished to introduce CFLs would have to compete fiercely for a small amount of additional grocery store shelf space, normally segregated from the rest of the lighting display area in a fluorescent-only section or in temporary end-cap displays. [24] This phenomenon was and is not limited to grocery stores. Philips was for many years the nearly exclusive provider of lamps to Home Depot as Osram Sylvania was to Lowe's and General Electric was to Wal-Mart. This meant that comparison shopping across CFL models and technologies had to be done store-to-store, rather than within a single store.

Manufacturers themselves recognized the problem. According to the *Home Energy* article, GE and other manufacturers noted that what will save the CFL is the fact that consumers are purchasing more of their lighting products in stores other than grocery stores, such as hardware, home improvement (Home Depot, Home Base), and discount general merchandise (Target, Wal-Mart) stores--and that they're willing to spend larger amounts of money on lighting products in this type of store. "As light bulbs become thought of less as a commodity like eggs and milk, and more as a home improvement purchase, price sensitivity will go down." [14] A report published in 1999 by the Natural Resources Defense Council (NRDC) [24] noted that "among the largest national chain retailers, niche CFL manufacturers could gain a foothold if they offered types of products not already provided by the full-line manufacturers. Indeed, smaller CFL manufacturers were able to garner substantial market share because of their ability to sell next to products made by General Electric, Osram Sylvania, or Philips in stores that would otherwise rarely carry products by two of the Big Three." [24] Smaller CFL manufacturers were also able to make advances through independent retailers and smaller franchised hardware stores like Ace and TruServ who do not have exclusivity agreements with the big three but purchase instead through regional distributors. Other non-traditional sales channels included mail order catalogs and Internet sites hosted by private companies and DOE. [24]

2.4. Manufacturer Barriers

Main manufacturer barriers include the market structure issues discussed above along with lack of profit motivation, lack of marketing, competition from other technologies, and fixture manufacturer issues.

2.4.1. Lack of Marketing

A report by the NRDC noted that before 1990, CFLs were really not marketed to residential customers at all; despite some isolated media demonstrations, they were nearly impossible to find in U.S. retail stores and instead were selling primarily to commercial and industrial customers through wholesale channels. Any residential customer who obtained a CFL likely got it through a utility program. At the beginning of the 1990s, major manufacturers began explicitly packaging CFLs to the residential consumers with full- color packaging instead of plain white boxes. [24]

However mass-market advertising was still lacking. "Consumer education is critical, yet no manufacturer promotes CFLs on radio or television" a *Home Energy* magazine article stated in November 1993. [10] A 1994 *Home Energy* article noted, "Forget educational campaigns; up until now, none of the major manufacturers have had any kind of advertising at all for CFLs aimed at the residential market."[14]

Manufacturers claimed that educating the consumer was difficult. "We're trying to educate the consumer," Dick Dowhan, manager of corporate communications for Osram Sylvania was quoted as saying in the 1994 *Home Energy* article, "but it's a very difficult sell." "The single biggest challenge facing the industry is education," said Gary Gumz of GE. [14] "You have to get the consumer to make a trial purchase." GE was the only major manufacturer planning a television commercial to support their bulbs according to the 1994 article. [14]

A 1999 study by Heschong Mahone for the California Energy Commission noted that news about the technological advances in CFLs had not widely reached the residential consumer because CFLs were not advertised as a consumer product in the mass media. "While the three major lamp companies do engage in competitive advertising to commercial customers, they have rarely used consumer based advertising for promotion of efficient lighting products in homes. Instead competition for market share among the lamp companies seems to be based more on competition for the loyalty of retail distribution outlets, rather than directly reaching out to the consumer about the merits of a particular product," the report noted. [20]

A 1999 NRDC report noted that utilities and energy efficiency organizations were spending over $23 million a year to try to shift $1 billion of consumer purchases from incandescent to compact fluorescent bulbs. "This is quite a challenge in a country where $100 million nationwide, TV-focused advertising campaigns are normally deployed to introduce new consumer products. Without more effective leveraging of advertising funding from manufacturers and retailers, this will continue to be an uphill battle," the report stated. [23]

That national marketing did not really kick in until the ENERGY STAR and "Change a Light, Change the World" programs of the early 2000s.

2.4.2. Other Technologies

A 1998 study for the Northwest Energy Efficiency Alliance noted that competition from other non- incandescent lighting products (such as the halogen infrared PAR lamp and high-performance metal- halide lamps) also contributed to the market's slow adoption of CFLs. [17] A Smithsonian article noted that GE shelved several early compact fluorescent designs in favor of the Electronic Halarc, a small metal halide lamp, and that its failure temporarily placed the company in a difficult competitive position. [1] While competing technologies is not an issue unique to residential lighting, it does point to some important lessons to be learned for those wishing to promote a specific technology, in particular the need for consumer education and clear marketing messages and terminology. For example how is the consumer to distinguish between several different lighting technologies all promoted under the name "energy efficient lights" or with vague monikers like "Earth Light"? [41]

2.4.3. Fixture Manufacturer Issues

While this report deals primarily with integrated screw-based CFLs, there is a growing market for lighting fixtures that use dedicated, pin-based CFLS with a separate ballast, thanks in part to new 2005 California Title 24 energy code requirements requiring dedicated energy-efficient lighting fixtures (or motion sensors) in most rooms of the house. And it can be argued that dedicated fluorescent fixtures can be economically, optically, and thermally optimized for use with fluorescent light sources, while screw-based CFLs are a compromise placed in fixtures intended for use with conventional incandescent sources, and requiring the user to discard both ballast and lamp every time one or the other fails.

In focus group studies throughout the 1990s fixture manufacturers identified barriers to CFL adoption not identified by integrated CFL manufacturers. They expressed concerns about the rapid change in CFL lamp designs and difficulties and costs in retooling their manufacturing lines to keep up with these changes. [16,20] They noted poor performing ballasts and as late as 1997 some fixture manufacturers said they were going back to the (less efficient) magnetic ballasts because of performance issues. [16] They expressed concern about foreign competitors copying their fixture designs and selling lower priced, inferior lamps that would turn off U.S. consumers to CFL fixtures due to their poor performance [16]. They were also concerned about the lack of standardization and compatibility in CFL lamp-ballast combinations and expressed fears that when it came time for consumers to find replacement lamps to fit the ballasts in their fixtures they would not be able to find compatible lamps. [16,20] Coordination and communication in the fluorescent lamp and fixture manufacturing industries and standards organizations could help address compatibility issues.

Lighting retailers who were interviewed in 1996 identified lack of CFL fixtures as a significant barrier to CFL adoption and noted that the few CFL or linear fluorescent fixtures that were available were very utilitarian looking. [16] A lighting design competition conducted by the American Lighting Association, the Consortium for Energy Efficiency, and DOE that was started in 2003 and has received positive media and enthusiastic participation by some lighting manufacturers is addressing this need.

2.5. Retailer Barriers

Significant retailer barriers to CFL market introduction were lack of awareness of and misinformation about the technology, which resulted in a reluctance to commit shelf space to CFLs.

Retailers employed a variety of different profitability targets. CFLs might be more profitable than incandescents on a dollars per million-lumen hours basis. However, retailers might conclude they were less profitable than incandescents on a percentage markup or "turns" basis (number of products sold per linear foot of shelf space per day).

Home improvement retailers were largely employing the "bazaar" or "flea market" business model at this time, whereby they competitively allocated selling space to various manufacturers and expected them to do much of the work of earning them a reasonable rate of return on that space. Slow performers would lose their allocations to other manufacturers. Thus retailers relied on manufacturer representatives to train their staff on how to educate consumers about the benefits of their products, set up point of purchase displays, and even maintain inventory on the shelves. Utilities increasingly found in the late 1990s that they needed to step in and assist with these functions through field staff to ensure that CFL sales would be brisk, since many of the new, niche CFL manufacturers that used utility incentive awards to secure retail allotments lacked the infrastructure to provide retailer training, inventory maintenance, and compelling point-of-purchase displays.

2.5.1. Lack of Awareness and Misinformation

A 1992 EPRI study of lighting retailers found that many saw it as a temporary technology that would be replaced by something else within the decade. [8] This study found chain store lighting department managers had many misconceptions about the technology and seemed to have no more technical knowledge than the average residential customer. [8]

A California Energy Commission 1997 study of retailers showed their understanding of energy efficient lighting was surprisingly rudimentary. When asked to name their best-selling energy-efficient products, one-third included standard incandescent lamps. If they did mention fluorescent lighting, they usually mentioned four-foot fluorescent tubes, not CFLs. [16]

This same 1997 California Energy Commission study of retailers showed a majority perceived fluorescent fixtures and lamps to have low light output and other objectionable characteristics such as poor light color and ballast hum. These attitudes reflect poor experiences with early models of CFLs. There seemed to be no awareness of the major changes in technology that had occurred within fluorescent lighting (i.e., electronic ballasts) to address these problems. Most of the products they were selling were older fluorescent technologies like tubes and circline lamps. "The [retailers'] lack of understanding of fluorescent technology would seem to be a major market barrier for residential consumers." [16]

Retailers acknowledged the problem. As early as 1991 retailers told EPRI that more advertising and consumer education was needed and that utilities should provide this to help break down the high first- cost barrier. They also wanted to see more programs from manufacturers. They praised utility coupon programs. [8]

In a 1999 report on the Northwest LightWise program, 60% of the retailers interviewed still did not feel they had enough information about CFLs to adequately sell the product. Overall they felt unprepared to explain CFL benefits, wattage conversions, power quality, and differences in ballasts. [21]

This lack of awareness and education on the part of retailers could be a major hindrance to furthering any new consumer technology. In a 1998 telephone survey of 500 Northwest utility customers, only 57% acknowledged that they had ever seen, read, or heard anything about CFLs and of those, 62% mentioned *retailers* as a primary source of information about the bulbs. [17] From 2,000 customer response cards turned in by customers who had purchased CFLs through the LightWise program in 1997, 65% said they had heard about LightWise from retailers, either from store displays or from store employees while only 23% mentioned utility bill inserts as their source of information. [17] This barrier has been addressed by utility programs that offered retailer training as key program components.

2.5.2. Lack of Shelf Space

An NRDC report published in 1999 noted that even in parts of the country where utilities had funded efficient residential lighting programs for nearly a decade, CFLs still occupied only 4% to 7% of the retail shelf space for household light bulbs, and less than that in grocery stores. [24] Even in 2006 many grocery, drug, and small hardware stores carry only a limited variety of bulbs (personal communication M. Ton, Ecos Consulting, 5/2//06).

CFLs were fighting ingrained retailer practices. For example, conventional bulbs are an important promotional item for hardware stores, especially during "light bulb season" (fall and winter) when many hardware stores give away free light bulbs to attract customers. [14]

A 1998 evaluation of the LightWise program for the Northwest Energy Efficiency Alliance found that many retailers did not stock CFLs on a year-round basis. Retailers did not see CFLs as a profitable product line and were hesitant to devote the necessary shelf space in their stores. [17]

A program evaluation of the Northwest LightWise Program by the Northwest Energy Efficiency Alliance found that, in 1996 among retailers who did carry CFLs, over 70% of the CFLs were stocked in the undesirable shelf location 2 to 3 feet above the floor or in dump bins. However utility programs could be effective in changing this. The LightWise Program evaluation found total lineal footage devoted to CFLs increased from 182 feet in January 1996 to 261 feet in May 1997 in Oregon and Washington. Over the three years covered in the program, LightWise products began to be placed in mid-aisle locations, areas where permanent products are normally stocked. [17]

2.6. Electrical and Building Contractors

Electrical and home building contractors can be significant influencers on CFL purchases both by the choices they make for lighting fixtures in new construction and the suggestions they make to homeowners doing remodels, so lack of awareness or enthusiasm on their part can be a barrier to CFL adoption. Studies have noted a reluctance among some in the home construction industry toward innovation. Fear of callbacks might be one reason. CFL fixture light quality, color, and amount of light were mentioned as reasons for customer complaints and callbacks by one-third of the 26 California residential contractors surveyed in a California Energy Commission 1997 study. [16] An NRDC 1999 report noted that utility program efforts to date to encourage major increases in electrical contractor CFL sales had been less well received than retail-based programs because contractors are generally constrained by very tight project budgets and derive no benefit from lower future energy bills. [24] Education aimed at this group could pay respectable dividends.

2.7. Other Barriers

EPRI pointed to the United State's segmented lighting industry, lack of product performance standards, and lack of coordination among manufacturers, utilities, and retailers as additional barriers to CFL market growth in the United States according to a 1993 *Home Energy* article. [10]

3.0. Drivers

There were several drivers for increased acceptance of CFLs because of their energy-efficiency benefits: utility programs, regional energy efficiency programs, nationwide efforts like ENERGY STAR, the impact of the Western electric energy crisis of 2001, and the impact of California's Title 24. These and other drivers are described in this section.

3.1. Utility Programs

In the early 1990s many utilities sponsored programs intended to drive consumers toward CFLs. As a 1990 PG&E study reported, 72% of participants said they would not have purchased CFLs without the program. [7] The goal of utility financial incentive programs (like giveaways, coupons, and rebates) is to temporarily lower the cost to whet the consumer's appetite for high-efficiency products, thus driving up demand which hopefully will lead to increased production, affording economies of scale in production and distribution, resulting in permanent price reductions, a process otherwise known as market transformation as shown in Figure 3.1 [23]

Many early programs were simply product giveaways intended to familiarize customers with CFLs by distributing free samples. Through the 1990s, utility programs evolved from incentives given directly to the consumer, to utility-run mail order programs, to incentives delivered through retailers via rebate and coupon programs, to incentives delivered to manufacturers, who passed greater savings on to consumers through discounted wholesale prices.

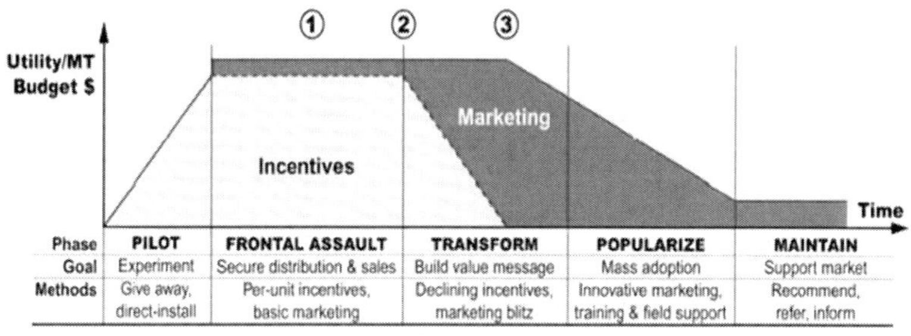

Figure 3.1. Evolution of a Utility Market Transformation Program [23].

In the late 1990's electric utilities substantially reduced their program budgets for energy-efficient residential lighting as part of broad-based cost-cutting in response to utility industry restructuring.

A 1999 NRDC report noted that utility energy-efficiency investments dropped by about $736 million between 1993 and 1997, and 42 utilities stopped funding efficiency programs entirely. [23] Several West Coast utilities stepped up funding of lighting programs again in response to the energy crisis of 2001 fueled by emergency state funding to minimize blackouts.

The pros and cons of the various types of utility programs that have been implemented are outlined in Table 3.1.

Industry consultants caution that utilities manage their expectations about controlling product flow. For example, under some buy-down scenarios, where products are shipped to distribution centers, manufacturers tend to have less control over timely shipment, availability, or distribution, which can result in missed promotional dates, (personal comm.. with My Ton, Ecos Consulting May 2, 2006).

3.1.1. Utility Give-Aways, Direct Installs, and Mail Order Programs

During the late 1980s and early 1990s, utility give-away, direct install, and discount mail-order programs put millions of CFLs into the hands and homes of consumers and helped many to try CFLs for the first time. CFLs were handed out at state fairs, included in new home packets by realtors, sold through school and civic group fundraisers, and distributed door to door by utilities. [9,11,13,29]

Some utilities conducted direct installation programs, where utility employees or contractors entered people's homes to install CFLs in the fixtures.

This guaranteed the CFLs were installed, which was an important issue for utilities attempting to ensure program quotas especially with early CFLs, which didn't fit many fixtures, and again after the West Coast energy crisis when California utilities were anxious to get participation to reduce load. Over time, giveaways became limited primarily to low-income programs, as utilities sought ways of encouraging most of their customers to make at least a partial investment in efficient lighting products on their own. Some utilities established mail order catalog programs with embedded utility incentives so customers could have qualifying products sent directly to their door. Utilities also set up programs aimed at the hotel and multifamily housing industries. Some utilities even leased CFLs to their customers, recouping payments on a monthly basis as a bill surcharge.

Table 3.1. Pros and Cons of Utility CFL Programs

CFL Delivery Mechanism	Pros	Cons	Notes
Manufacturer Buydown	• Offers better control over pricing, distribution, and tracking • Minimizes over-subscription • Can stipulate 'sell-through' rates in agreements with manufacturers and retailers • Also offers utility an opportunity to promote selected CFLs	• Less customer recognition of utility efforts. • Can be unevenly paced • Difficult to target end customers • Can be disruptive in more mature markets • Less opportunities for customer education/outreach coordination	• Examples include BPA's 2005 Fall Campaign, early Alliance and regional efforts, SCE's 1995/6 programs • *Costs:* can be low, $0.015 or less per lifetime kWh, depending on buy-down level
Customer Coupons	• Stimulates sales through implied cash value • Exclusivity- customers feel they are getting a deal • Customer recognition of utility effort is higher with coupons • Coupon expirations create urgency to buy now rather than later • Provides more opportunities for promotions/education/outreach	• Better suited for regional/larger • programs where volumes justify • marketing/education efforts • Utility-to-ratepayer direct distribution has traditionally shown low response • Redemption rates do not match buydown sell-through rates • Additional administrative costs • Less buy-in and coordination with manufacturers and retailers	• Examples include BPA's 2001 Campaign, IPCo 2004, and PSE's 2002 – 2003, and 2005 Programs • Costs: can also be low, ranging from $0.014 (PSE 2004) to $0.021 (BPA 2001) per lifetime kWh

CFL Delivery Mechanism	Pros	Cons	Notes
Direct Install	• Better control over target market • Can use existing infrastructure • "Guaranteed" installation • Installers can be trained to target "best" sockets/applications	• Need additional infrastructure • (contractors/call centers) • Better suited for regional/larger programs due to infrastructure required	• Successful examples include ETO 2002-2005 • Costs: ETO's cost is about $0.03 per lifetime kWh
	• Also offers utility an opportunity to promote only the best CFLs • High customer recognition of utility efforts • Good educational opportunities	• Additional administrative costs, depending on utility • May not work for rural areas	•
Bundled Delivery	• Provides additional opportunity for savings • Can directly target market segments (multi-family, low-income, etc.) • Installers can be trained for best CFL applications • High customer recognition/satisfaction • Good educational opportunities	• Requires infrastructure set up (contractors, call center, incentive processing) • May require additional administrative cost, depending on program design • Marketing cost can vary, depending on target market (rural, urban)	• Successful examples include ETO's 2002 – 2005 • Costs: ETO's cost is about $0.017 (multi-family) to $0.022 (single family). Cost includes CFL savings only.

1. Program costs depend heavily on design and ultimate goal (i.e. market transformation or kWh acquisition), as well as length of program, and program metric of cost effectiveness (TRC, levelized cost, lifetime kWh costs)

Bundled programs offer more opportunities for savings beyond CFLs, but requires more design and infrastructure support.

While give-away programs did increase local awareness and temporarily reduced the price barrier, subsequent program evaluations noted several drawbacks. Many of the CFLs initially distributed were low-cost, low-quality lamps. Chosen in an effort to keep the unit price low, these poor-performing demonstration lamps may have only reinforced negative attitudes. [20] As an NRDC report noted “It is difficult to convince consumers to purchase something that was once given to them for free and that did not provide a positive experience.” [24] Direct install rates did not guarantee retention or consumer acceptance and side-stepped the process of customer choice altogether. “At least one direct-install program in New England in the early 1990s experienced removal rates of over 30% within the first year of installation.”[24] (However, it should be noted that customer satisfaction rates with CFLs did go up as the technology improved [31]) A Seattle City Light report noted the nearly universal opinion of conservation professionals that, in contrast to unsolicited give-aways, programs that required customers to take some action to receive CFL bulbs would have a much higher installation rate and would strengthen customers’ knowledge of and commitment to the technology. [31,32] Another danger with give-away programs and utility direct sales programs is their potential to undermine sales of CFLs at retail stores. [11]

3.1.2. Retailer Rebate and Coupon Programs

Many utilities mailed coupons to their customers as utility bill stuffers, encouraging the customers to redeem them at participating retailers for discounts off of qualifying products. In an effort to further reduce customer hassle, some utilities made arrangements with retailers to offer price reductions with in- store coupons and then reimbursed the retailers for discounts after the sale. These programs brought purchase prices down considerably for customers. For example, through a 1994 Central Maine Power program customers could buy CFLs for $3.99 that had retailed for $19 to $24. Of the coupon and rebate programs, point-of-purchase coupons had nearly 100% redemption rates, while mail-in rebates had redemption rates of 20% to 80%, and coupons mailed to customers’ homes had similar redemption rates. [11] However, retailers were sometimes reluctant to participate in instant rebate programs due to the paperwork involved. [13] The enthusiastic and on-going participation of retailers is one of the building blocks of the program, and one that needs to be acknowledged sincerely and frequently, as a 1999 program evaluation of the Northwest Energy Efficiency Alliance’ LightWise program evaluation report noted. [17]

3.1.3. Incentives to Manufacturers

Several utilities switched to programs that gave the financial incentive directly to the manufacturer when they realized the larger savings to the consumer and to themselves in terms of lower program costs. [10,17,18,22,24,28] In 1992, Southern California Edison switched its $5 per lamp rebate from the customer to the manufacturer and found much better results at lower program costs. For example if the wholesale cost of a given CFL was $10, and the typical retail markup was 67% ($6.70), the lamp would retail for $16.70. A $5 customer rebate would lower the retail price to $11.70 but a $5 manufacturer rebate would cut the wholesale price to $5. With a 67% markup ($3.35), the retail price would only go up to $8.35. SCE gave manufacturers additional incentives for higher power factor, good color rendering, and high lumen output and also for offering additional discounts and cooperative advertising money for their retailers. SCE gained from lower administrative costs – they went from 70% of program funds for the consumer rebate program to 10% of program funds for the manufacturer program. In 1992, almost 1 million CFLs were sold under the program and annual electricity savings amounted to 53 million kWh. SCE repeated the program in 1993 and 1994 and it became the model for larger regional and national programs thereafter. [11]

3.1.4. Niche Marketing – Multifamily, Elderly, Urban, Ethnic, and Rural

Multifamily housing and senior citizens were two markets perceived to be drivers for energy efficiency in residential fixtures identified by fixture manufacturers in a 1997 CEC study. Apartment owners who foot the bill for outdoor and common area lighting would see the gains of large-scale replacement of incandescents with energy-efficient lighting. It was thought that older consumers would be interested in the inherent safety that longer lasting bulbs would provide by requiring less frequent trips up a ladder to do light bulb changes, especially with fixtures located on high ceilings. [16]

The hotel industry was also targeted by DOE and some utility programs. Hotels could save significantly on their lighting bill by replacing incandescents with CFL in common areas and guest rooms. In addition to energy savings, hotels also benefit from lower maintenance costs due to less frequent lamp failures. Large orders for CFLs from buyers like hotels and multi-family housing helped support increased CFL production and economies of scale.

San Diego Gas & Electric (SDG&E) and Southern California Edison (SCE) were two utilities who joined forces in a program specifically aimed at hard-to-reach residential customers in rural hardware stores and ethnic grocery markets that serve non-English-speaking customers (Hispanic, Chinese, Vietnamese, and Korean). They hired Ecos Consulting to run their 2002-04 program which offered a $5 per CFL incentive to manufacturers, distributors, and retailers participating in the program. Ecos also provided marketing materials and one-on-one support to retailers and manufacturers through program field representatives. The program resulted in participation by dozens of retailers in the service territories. Most manufacturers praised the program and said they would participate again in similar programs, although they cautioned against confusion caused by multiple programs (there was a separate state-wide program in effect at the same time aimed at larger retailers.) Lessons learned included that programs with this audience take at least two or three years to build trust among the retailers. One-on-one education of retailers through field representatives was crucial to build trust and to educate retailers and customers about the value of the new technology and the difference in quality between ENERGY STAR and non-ENERGY STAR products. There is a need for consistency among programs. Having multiple programs operating at the same time with different incentives made program delivery difficult and engendered confusion and resistance among affected customers. [34]

Several studies noted the slow adoption of CFLs in rural America and throughout the South, due to the lack of utility programs in those areas and, especially in rural areas, to the lack of competition due to the limited number of retailers, limited shelf space, and exclusivity agreements between retailers and large bulb manufacturers. The increase in the availability of CFLs to rural areas in the early 2000s was due notably to the expansion of large home improvement center chains and Wal-Mart.

3.2. Regional Efforts

As early as 1989, utilities began joining forces to develop regional energy efficiency programs, including CFL programs. In 1989 two groups of Northwest utilities formed with support from Bonneville Power Administration and the Washington State Energy Office: the Northwest Residential Efficient Lighting Group and the Northwest Residential Efficient Appliance Group. These later combined and changed their name to the Residential Efficient Appliance and Lighting group (REAL). [11] In 1991, Pacific Gas and Electric (PG&E) worked with the NRDC and other utilities to form the California Compact, a non-profit corporation dedicated to increasing consumer awareness of efficient lighting and increasing the availability and affordability of CFLs. [11] In late 1991, the Western Utility Consortium was formed among utilities in California and other West Coast states, especially for those whose demand-side management (DSM) programs involved trade allies such as appliance and lighting product retailers and manufacturers. [11] Also formed in 1991 was the Consortium for Energy Efficiency (CEE), a non-profit, public benefit corporation that promotes energy-efficient products with members including electric, gas, and water utilities; research and development organizations; and state energy offices; with support from the U.S. Environmental Protection Agency (EPA) and the U.S. Department of Energy (DOE). [16]

The LightWise Program was started in 1996 by the Northwest Energy Efficiency Alliance, representing 120 public and private electric utilities, to accelerate awareness of CFLs. [17] In the Northeast, Northern Lights and then the Northeast Energy Efficiency Partnership (NEEP) were formed to coordinate efficiency activities. Wisconsin Energy Conservation Corporation (WECC) developed a similarly uniform program concept across multiple utilities in Wisconsin. At the same time, energy officials from eight states in the upper Midwest (Minnesota, Wisconsin, Illinois, Michigan, Indiana, Ohio, Iowa and Missouri) launched the Midwest Energy Efficiency Alliance (MEEA) to formulate new efficiency initiatives for the region. [24]

Regional programs worked with and learned from manufacturers in developing these programs. For example, Lightwise changed its requirements on power factor after losing major manufacturer participants. Lightwise initially required manufacturers who wished to be in the program to provide bulbs with a power factor of 0.9 or higher. They lost Osram Sylvania as a participant in 1998 after Osram Sylvania decided it would no longer produce high power factor bulbs, which it perceived as a "niche" market. [21]

These regional groups sponsored conferences, coordinated programs, and conducted promotional activities to encourage the use of CFLs. A 2003 article by Platts, a consulting firm for utilities, highlighted one of the roles they can play. "Retail outlets typically provide space for the products that they carry, but little else in the way of support. Whether it's breakfast cereals or CFLs, it's up to the manufacturers or some other party to provide any extra promotion that a product may need. The California Residential Lighting and Appliance Program (CRLAP) and the Northwest Energy Efficiency Alliance's programs both used field services teams that included staff and subcontractors to play that role. Typical activities included product merchandising— fashioning product displays, providing end-cap displays, verifying product pricing, ensuring product availability, and generally making the products more visible to shoppers—as well as employee training, on-site promotions, and reporting and responding to signals in the marketplace. In 2002, outreach activities for the Alliance's program handled by field services teams included 37 training sessions, 102 on-site promotions, and 2,349 retailer visits." [30]

"The relative cost of participating in regional and national collaborative forums is small," a 1994 Home Energy article summarized, "but the potential effectiveness of such efforts in simplifying and leveraging the response of trade allies is enormous." [11] Now that these regional efforts are in place, a new lighting technology with strong energy-saving potential is likely to benefit from regional support and coordinated approaches to promotional programs, versus a variety of approaches introduced by individual utilities.

3.3. Foreign Competition

According to a 1994 *Home Energy* article, "Resistance to change by the big 3 manufacturers may have been the policy in the past, but the U.S. market is not a closed system, and U.S. manufacturers can't ignore pressure from outside." The article went on to discuss Panasonic, a small player in the U.S. lighting market, but a division of Matsushita Electric Corporation, which controlled a large portion of the lighting market worldwide and was planning a significant U.S. launch in 1994. [14] Panasonic achieved market penetration with a range of other energy-efficient products, but made limited inroads in the US the CFL market, pinned between the three largest full-line lamp manufacturers on one side and the CFL-only niche manufacturers on the other. Foreign owned and smaller U.S. manufacturers were eager to participate in government-sponsored CFL procurements in the late 1990s. [19]

In 2001, the European Union imposed steep tariffs (up to 75%) on Asian manufacturers in response to charges of dumping (below-cost selling) of CFLs. This caused numerous suppliers from Asia to shift their marketing efforts to North America, greatly increasing the available supply of CFLs, especially low-cost products, and driving additional price competition. [28]

3.4. California Title 24

California has traditionally had some of the most stringent minimum building energy efficiency requirements in the nation. California's Title 24 Building Energy Standards, instituted in 1978, required that efficient lighting fixtures be installed in the kitchen and bathrooms of new homes. While California's Title 24 deals primarily with pin-based lighting, not screw-based CFLs, it is an interesting case study of the impact of legislated requirements on lighting choices. And, due to the sheer size of the California residential market, revisions to Title 24 enacted in October 2005 (described in Section 6 below) requiring dedicated, non-screw-based energy-efficient lighting (or controls) in every room of the house will undoubtedly influence the market, at least in terms of increasing fixture availability, competition, and attention to CFL design by manufacturers.

In a 1997 survey of 40 national manufacturers of CFL and standard residential lighting fixtures conducted by Heschong Mahone Group for the California Energy Commission, California's Title 24 energy code was acknowledged as one of the most important influences in the national fixture marketplace. Some manufacturers said it was the biggest single impact on the market. They felt it had accelerated development of the electronic ballast and forced the lighting industry to be more creative. [16] The survey of fixture manufacturers found that "while Title 24 was mentioned 22 times as a major influence on energy efficiency in lighting products, the Green Lights program was mentioned only once. Utility DSM programs were mentioned four times, always in the past tense. They were seen as a brief perturbation in the market, sometimes as a positive influence, but just as often as a negative influence. Manufacturers rightly do not want to be subject to multiple, conflicting standards or frequent or erratic changes in the market. Stable, long-term horizons allow manufacturers to plan for the future and make more secure investments. Appliance standards, especially those that are coordinated at a national level and phased in over a multi-year period, provide that stability for wise investments." [20] A survey of California residential contractors in 1997 found most were installing fluorescent fixtures most of the time, even though half their jobs were remodels (not new construction) and therefore were not required to follow Title 24. [16]

3.5. Western Electricity Crisis of 2001

The crisis was due to several factors: California's early foray into electric utility deregulation, which removed the safety cap allowing rates to skyrocket; a hot summer of 1999 followed by a cold winter, which together with a booming economy greatly increased demand; high natural gas prices; and droughts and cold weather in the Northwest, which diminished power available in the Northwest for export to California. All of these factors set the stage for a crisis that got worse, not better as 2001 progressed, with episodic power shortages, rolling blackouts, and California's largest utility, PG&E declaring bankruptcy. California experienced 70 days of Stage 1 emergencies (voluntary conservation requested), 65 days of Stage 2 emergencies (interruptible customers curtailed), and 38 days of Stage 3 emergencies (involuntary rolling blackouts) in 2001. [28]

In response to these events, California's state and local governments mobilized a massive campaign to cut consumption. The California legislature authorized $20 million in emergency funding for the California Conservation Corps (CCC) to purchase and distribute 1.9 million CFLs and 1.4 million "Flex Your Power" brochures to low-income households in every county in the state. Elected officials made public service announcements urging customers to conserve. The state convinced nearly a third of California's households to participate in the state's "20/20" program. This granted 20% electric bill rebates to households that reduced their overall summer electric use by 20% compared to the previous year. [28] Rising electricity rates, especially for usage above baseline amounts, gave consumers a powerful new motivation to cut demand. While baseline electric rates held steady, Californians saw very steep increases in their incremental rates for consumption above baseline. (www.sce.com/NR/sc3/tm2/pdf/ce12-12.pdf). [28]

These events drove up CFL demand and, together with the sudden appearance of low-cost Asian CFLs, sparked unprecedented competition among manufacturers and retailers to gain or at least maintain market share. This apparently resulted in substantial internal discounting and created a "virtuous cycle" in which "price cuts drove greater sales, increasing economies of scale in manufacturing and making possible even lower prices on subsequent orders. These competitive forces may well have done more to reduce average bulb prices than utility incentives, especially outside of California and the Pacific Northwest." [28]

"One of the clear lessons from 2001 is that disruptions in electricity supply (and the resulting inconvenience, higher rates, and saturated media coverage) can drive the same kinds of short-term improvements in product efficiency that the oil crises of 1973 and 1979 did with vehicles. However, such responses are, by themselves, inherently transitory, and consumer behavior can be expected to revert back to previous levels if not sustained by other factors." [28]

Other factors were at play though; marketing efforts were finally moving from local and regional utility- sponsored efforts to a national movement with the ENERGY STAR program.

3.6. DOE CFL Programs

In March 1998, DOE announced that Link USA, Lights of America and Sunpark Electronics had won a competitive procurement conducted by PNNL for DOE to promote the introduction and early sales of subcompact CFLs (sub-CFLs). These lamps were no longer than 5.75 inches, had a screw-in base and provided significant energy-efficiency benefits, met high performance requirements, and came with a one year manufacturer warranty. A second phase of the program in 1999 resulted in additional sub-CFL models produced by Phase 1 manufacturers as well as new models offered by JKRL and Surya. These smaller sub-CFLs met high technical specifications for color rendition and light output, and were comparable in length to ordinary incandescent bulbs. Sub-CFL sales were promoted to large-volume buyers primarily through the web site www.pnl.gov/cfl. DOE's program succeeded in introducing 17 new sub-CFL models to the market at volume prices (including delivery) ranging from $5.75 to $9.61. By the program's cessation in May 2001, over 3.3 million sub-CFLs had been sold. DOE transitioned the program to the Northwest Energy Efficiency Alliance (NEEA) who changed the name to BetterBulbsDirect.com and promoted many of these same lamps to small- and medium-sized retailers as a way for them to access high-quality, competitively priced sub-CFLs. Many of the Sub-CFL Program performance specifications were subsequently adopted by DOE as the first ENERGY STAR specification for CFLs.

DOE has supported other CFL development projects, for example the Berkeley table lamp project, the ballast-in-a-socket device, and development of electrodeless lamps through R&D cooperative agreements and support of national laboratory research. [26].

3.7. ENERGY STAR

ENERGY STAR®, an energy-efficient product labeling program sponsored by U.S. Department of Energy (DOE) and the U.S. Environmental Protection Agency, launched a program for residential light fixtures in 1997.

This brought a clearly recognizable national brand to a fragmented marketplace. It gave utilities and regional energy efficiency groups a benchmark of lighting performance and quality around which to rally their marketing strategies. This program inaugurated a number of high-visibility marketplace efforts, including the compact fluorescent torchiere. The establishment of the ENERGY STAR (only need to have the registration mark once in a document) fixture program and the DOE-Sub- CFL Technology Procurement caused many utilities and regional organizations to argue for the establishment of an ENERGY STAR labeling program for screw-base CFLs as well. This program was launched by the DOE in May 1999 and garnered participation by more than a dozen of the major CFL manufacturers and numerous utilities. [28] ENERGY STAR offered national branding and a single set of specifications for manufacturers to design to. [24]

Prior to the 1999 introduction of the ENERGY STAR CFL Program, CFLs were required only to meet certain reference standards for performance, safety, and labeling by the American National Standards Institute (ANSI), Illuminating Engineering Society of North America (IESNA), and Underwriters Laboratory (UL). Market share of CFLs nationally was less than 1%, but a number of new products were entering the market from unknown manufacturers and of questionable performance. Consumers were unfamiliar with CFLs and had no way of knowing how to distinguish between better performing lamps and poor performing ones.

DOE's motivation for introducing the ENERGY STAR CFL program was to "provide specific parameters and direction" to encourage the production of "high quality, highly efficient CFLs." The program included minimum product performance specifications and outlined what manufacturers needed to do to become an ENERGY STAR partner (see Table 3.2). Since 1999, the CFL qualified product list has grown substantially. At the beginning of 2001, the program included 17 manufacturers offering 161 qualifying products. By the end of the year, that list had grown to include 94 manufacturers and 455 qualifying products. [28] However, the initial 1999 ENERGY STAR specification and 2001 revisions allowed manufacturers to do their own product testing at their internal NVLAP accredited laboratory that they met the requirements. Complaints about product performance led to changes in the specifications as shown in Table 3.1. For example, DOE added new testing requirements including the 1,000-hour rapid cycle test, and life testing. These new requirements were developed in close coordination with industry and stakeholder groups in an effort to improve the overall quality and performance of ENERGY STAR qualified CFLs.

3.7.1. Program for the Evaluation and Analysis of Residential Lighting (PEARL)

In January 1999, *Consumer Reports* published a critical product evaluation saying that many CFLs failed to live up to their claims of product performance and longevity. Home Depot, Costco, and other retailers imposed a number of testing requirements and ultimately made changes in their list of approved vendors, heightening aspects of competition well beyond price. In response, DOE, EPA, NRDC, and a number of utilities worked together to establish PEARL. PEARL performs formal off the shelf random testing obtaining samples from a number of retail locations around the country and sending them to the Lighting Research Center (at Rensselaer Polytechnic Institute in New York) for testing.

Two rounds of testing identified a number of ENERGY STAR qualified CFLs that, based on PEARL testing, did not meet program requirements. DOE acted on these findings and in some cases disqualifiednon-compliant CFLs. (the ENERGY STAR criteria has always required testing at a NVLAP accredited facility – not an independent laboratory. The independent laboratory is a requirement of the Third Party Testing and Verification Program. [28]

3.7.2. Revised ENERGY STAR Requirements

The bar for ENERGY STAR qualification has been raised a number of times by DOE. Qualified CFLs must now meet more stringent performance requirements, and the program has moved from accepting initial qualification at 1,000-hours of rated life to requiring at least 40% of rated lifetime test information for ENERGY STAR qualification. In addition, DOE will require manufacturers to assist in funding independent, third-party testing to help provide quality control in the marketplace. DOE's goal is to work cooperatively with manufacturers and stakeholders to continue to drive the market to offer the highest quality products and to maintain the integrity of the ENERGY STAR label. Table 3.3 provides an overview of how some of the key ENERGY STAR performance requirements have been updated (made more stringent) over the years.

Table 3.2. ENERGY STAR CFL Program Minimum Qualification Criteria

	1999	2001	2003	2005
Certification	All photometric and lifetime tests to be done at a Nat'l Voluntary Lab. Accreditation Program accredited facility.	All photometric and lifetime tests to be done at a Nat'l Voluntary Lab. Accreditation Program accredited facility. Three steps to qualification: initial, interim, and final	Initial qualification increased from 1,000-hours to 40% of rated lifetime. Two steps to qualification: initial and final qualification.	Two steps to qualification: initial and final qualification. Required participation within the Third Party Testing and Verification program.
PEARL or Third Party Testing	none	PEARL – off the shelf testing of efficacy, power factor, CRI, CCT, lumen maintenance	PEARL – off the shelf testing of efficacy, CRI, CCT, power factor, run-up time, 1,000-hour lumen maintenance, lumen maintenance at 40% of rated life,, rapid cycle stress test, interim life testing Third Party Testing and Verification program established as a program requirement	PEARL - Same as 2003 Additional mandatory test cycle completed for elevated temperature life testing for CFL reflector products Continuation of Third Party Testing and Verification Program development as part of the ENERGY STAR criteria for CFLs. Program slated to go into effect on 1/1/07.
Efficacy (lumens/watt)	depends on wattage, see Table 3.2			Increased, see Table 3.2

	1999	2001	2003	2005
Correlated Color Temperature (CCT)	Between 2700K and 3000K. If not, packaging should clearly state the temperature and color of product (cool or warm).	Same	Same	Manufacturer must identify products as one of these correlated color temperatures: 2700K, 3000K, 3500K, 4100K, 5000K, or 6500K,(a)
Lumen maintenance	40%. Average lumen levels of 10 sample lamps must be > 80% of initial (100 hour) rating at 40% of model's rated life	Same, plus must meet or exceed 90% lumen maintenance at 1,000 hours. Prequalification	Same	Proposed changes are *1,000 Hours* Lumen Maintenance and lumen maintenance at 40% of Rated Life(b)
Average rated lifetime test	Manufacturer declared length of time at which 50% of tested lamps (typically 10) reach end of their individual lives	Same plus the addition of Interim Lifetime Testing (see below)	Same;	Same
Rapid CycleStress Test	None	Must meet 5 out of 6 cycles	Same;	Same
Interim Life Test	None	Must now have at least 8 out of 10 samples burning at 40% of	Same but included in initial qualification (in	same

Table 3.2. (Continued)

	1999	2001	2003	2005
		rated life; 2 failures= need justification, 3 failures=immediate delisting	2001, was required test during interim test).	
Warranty (residential use)	1 year OR a 1-800 number for consumers	At least 1 year AND a 1-800 number or web site for consumers Adherence to Federal Trade	At least 2 years for residential use; at least 1 year for commercial use	same
		Commission (FTC) labeling requirements	Labeling requires 800 number or web site, adherence to FTC labeling requirements	
Elev. temp. life testing				Suggested: elevated temp. testing for rec. downlights & indoor reflectors

(a) 9 out of the 10 samples tested must fall within a seven-step ANSI MacAdam ellipse for that color temperature at 100-hour lumen measurement. And **90%** of ongoing production (*as represented by samples tested from each production shift for same color when typically evaluated over 12 month period*) must fall within the seven-step MacAdam color ellipse associated with the designated (manufacturer declared) target color.

(b) Proposed changes are: **1,000 Hours:** Average lumen output of 10 lamps tested must be greater than 90.0% of initial (100-hour) lumen output @ 1,000 hours of rated life, and no more than two individual samples can have a lumen output measurement less than 85.0%.

Lumen Maintenance at 40% of Rated Life: Average of the 10 samples tested must be greater than 80.0% of initial (100-hour) rating at 40% of model's rated life (Per ANSI C78.5, Clause 4.10), and no more than three individual samples can have a lumen output less than 75.0%.

In August 2005, DOE released draft criteria that included the addition of specific test requirements for reflector CFLs identified for use within recessed downlights and indoor reflectors and increased efficacy requirements for bare and covered CFLs. The recessed downlights and indoor reflector CFL requirements included a short-term in-situ reflector testing procedure to determine initial luminous flux. The initial elevated temperature light output test procedure involves a two-step process. Lamps will initially be operated in a nominal 6” aperture, 8” deep insulated ceiling airtight (ICAT) downlight installed in the UL 1598 thermal test apparatus for IC-rated luminaries, less the insulation. This establishes the baseline illuminance corresponding to the light output determined under LM-66-00. Lamps will then be operated in the same apparatus with the insulation added. The ratio of the illuminance with insulation to the illuminance without insulation, expressed as a percentage, is Initial Elevated Temperature Light Output. The recessed downlights and indoor reflector CFLs will also be required to complete initial and final qualifications. Initial qualification includes elevated temperature testing at 1,000 hours of lumen maintenance, elevated temperature lumen maintenance at 40% of rated life, and elevated interim life testing. The final qualification is elevated temperature life testing.

See information on the revision to the ENERGY STAR criteria for CFLs at http://www.energystar.gov/index.cfm?c=revisions.cfls_spec

Table 3.3. Comparison of Minimum Efficacy Requirements for Past, Current, and Proposed ENERGY STAR Criteria for CFLs

Lamp Power (Watts) & Configuration	***Minimum Efficacy*: Lumens/watt (Based upon initial lumen data2)**	
	1999, 2001, and 2003 Criteria	**Proposed 2005 Criteria**
Non-specialty, Bare lamp:		
Lamp power < 10		**50.0**
10 ≤ Lamp power < 15	45.0	**55.0**
15 ≤ Lamp power ≤ 25	60.0	**60.0**
Lamp power > 25		**65.0**
Specialty, bare lamp (3-way, dimmable):		
Lamp power < 15		**50.0**
15 ≤ Lamp power < 25		**55.0**
Lamp power ≥ 25		**60.0**

Table 3.3. (Continued)

Lamp Power (Watts) & Configuration	*Minimum Efficacy*: Lumens/watt (Based upon initial lumen data2)	
	1999, 2001, and 2003 Criteria	**Proposed 2005 Criteria**
Covered lamp (no reflector):		
Lamp power < 15	40.0	**45.0**
15 < lamp power < 1919<Lamp power >	48.0	**50.0**
25	50.0	**50.0**
Lamp power ≥ 25	55.0	**55.0**
W/Reflector:		
Lamp power < 20	33.0	33.0
Lamp power ≥ 20	40.0	40.0

3.7.3. National Marketing: Change a Light Change the World

In the fall of 2001, Change a Light, Change the World, a national media campaign was launched with sponsorship from EPA and other sponsors. According to Calwell, more than 100 utilities, hundreds of retailers, and 25 manufacturers participated in the promotion, including many who had not participated in previous ENERGY STAR® promotions. Over 5 million radio listeners heard radio stories about Change a Light, while radio, TV, and print advertising reached tens of millions more. Calwell credits the national campaign with driving "a great deal of additional purchasing activity" despite the intense competition for news coverage from the events of September 11, 2001. According to Calwell, GE's advertising ran in 16 major papers throughout the country and achieved 26 million impressions. Its sales during the Change a Light timeframe were up by approximately 300% from the previous year." [28]

3.7.4. ENERGY STAR Participation and Lessons Learned

Before DOE established the ENERGY STAR CFL Program, a number of utilities and energy efficiency groups operated their own CFL promotional programs; for example, the LightWise program supported by the Northwest Energy Efficiency Alliance operated from 1997 to 2000. A key contributor to the success of the ENERGY STAR program has been the fact that utilities and energy efficiency groups embraced the criteria and outreach programs established by ENERGY STAR. These groups have provided valuable input for improving the specifications. This mutual cooperation provided many benefits for all concerned:

- Allowed efficiency groups to support common performance criteria, thus reducing efforts to establish such criteria on their own.
- Enabled efficiency groups to leverage resources through a national promotional campaign with ready-made public service announcements, advertising, and support materials.
- Offered manufacturers one set of criteria versus many, depending on where the products were sold.
- Provided ENERGY STAR with an interested and engaged group of organizations that would help define specifications and offer programs that encouraged the sale of ENERGY STAR CFLs.

Southern California Edison, Pacific Gas and Electric, the Midwest Energy Efficiency Alliance, and the Consortium for Energy Efficiency have all built successful demand side management and energy efficiency programs in support of ENERGY STAR CFLs.

There are several lessons to be learned from ENERGY STAR:

- Acceptance of common product performance criteria and manufacturer participation requirements (i.e., reporting, use of logos, etc.), like ENERGY STAR, can be important to the success of a new product.
- Third-party testing, as in PEARL, can also serve an important role in cases where product performance and compliance with specifications is an issue.
- The opportunity to participate in a national promotional campaign leverages the marketing budgets of many smaller local and regional entities and presents the public with a common theme and message and brand identification.
- Establishment or identification of an organization, like ENERGY STAR, that works with industry and stakeholders to establish product performance standards, monitors compliance with those standards, and provides labeling and educational support for compliant products can greatly improve the success of the new technology category in the long run.

3.8. CFL Sales Peak in 2001

After years of steady but unremarkable sales increases, CFL sales jumped in the United States in 2001. Relative to 2000 levels, national retail sales more than doubled, California sales increased by more than four times, and Northwest sales increased by roughly 10 times. National sales of CFLs grew from a 0.5% share of the screw-based lamp market in mid-2000 to a 1.6% share in mid-2001 and reached 2.1% by the fourth quarter of 2001 (Figure 3.2).

Calwell reports “Even accounting for sharp declines in the average selling price (ASP) of compact fluorescents, it is still interesting to note how much more dramatic the CFL market share became on a dollars basis. Remarkably enough, for the full year of 2001, CFLs outsold incandescent bulbs on a dollars basis in California.” Calwell notes that these figures indicate CFLs’ newfound importance to retailers as a source of revenues and profits. [28]

ECONorthwest reported Northwest sales totaling 8.6 million CFLs in the five quarters from fourth quarter 2000 through fourth quarter 2001, 20 times higher than the Northwest Energy Efficiency Alliance had predicted in their preprogram planning estimates. [29]

These higher sales in the west and northwest are attributable to the many factors mentioned above:

- power shortages in California and impacts on nearby regions
- deregulation and rising electricity rates
- improvements in CFL technology and testing for product quality
- lower wholesale CFL prices, due to economies of scale and greater competition
- expanded ENERGY STAR® promotions, including "Change a Light, Change the World”
- increased utility incentives and promotion in response to the energy crisis
- long-standing regional utility-funded efforts to build retail sales infrastructure and awareness. [28]

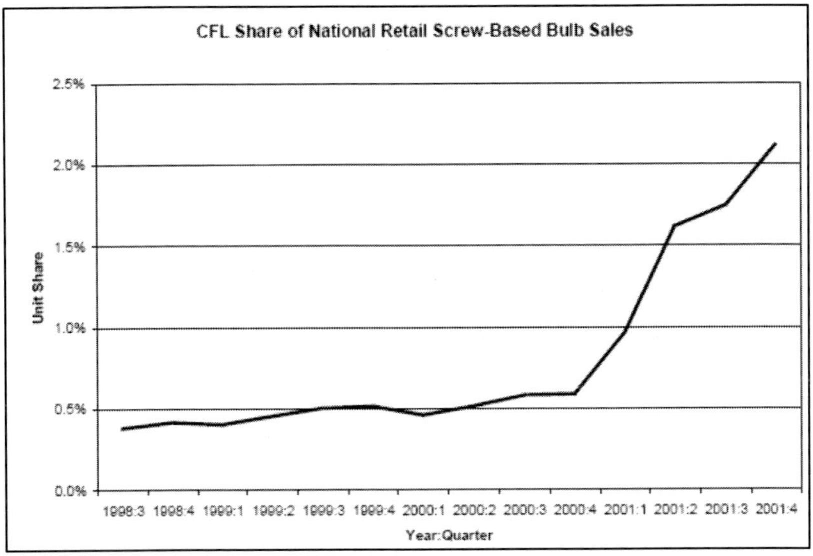

Source: RER 2002

Figure 3.2. National CFL Market Share, Third Quarter 1998 to Fourth Quarter 2001. (Source is Residential Market Share Tracking: Lamps. Prepared by Regional Economic Research for Southern California Edison. April 30, 2002, as cited in [28]).

These figures show that CFLs had been well positioned to respond to the energy crisis of 2001 but sustaining sales and increasing market share would not happen automatically. It would take continued marketing and refinement of the message to reach those consumers still not yet convinced. Section Four discusses several marketing approaches and themes applied or proposed through two decades of CFL development and promotion.

4.0. Marketing Messages, Tactics, and Trends – Suggestions from the Field

As we have seen in our discussion of barriers, technology improvements alone do not guarantee market acceptance. Getting the word out about the new product, showing consumers its value, educating retailers on how to market it, working with manufacturers to leverage promotion efforts– all of these are critical pieces of the consumer acceptance pie. Marketing activities and strategies have evolved as improvements were made to the CFL technology, as market players became more familiar with the technology, and in response to outside influences like the Energy Crisis of 2001, foreign competition, and the rise of do-it yourself (DIY) home improvement centers. Table 4.1 summarizes some of the major marketing trends. These are further discussed below.

4.1. Key Marketing Messages

Although CFLs were first introduced in 1979, marketing of CFLs to residential consumers didn't really start until around 1990; at that time only about one-third of the sales of screw-based CFLS were through retail stores while the remainder were through commercial and industrial distribution channels like electrical wholesalers. [28] A significant percentage of CFL distribution to residential users was via utilities. Many of the barriers in terms of availability, lack of advertising, and lack of consumer and retailer education have been identified above in Section 2.

Table 4.1. Marketing Trends

Trend	1980s-1990s	2000s
Key Messages	Talk about technology	Talk about benefits to consumers: longer life, cool and safe, less bulb changes, $ savings
	Like an incandescent	Better than an incandescent
Source for Bulbs	Utilities (giveaways, direct installs, bill stuffers)	Retailers (point of purchase displays, retailer training)
Utility Incentives	Given directly to consumer	Given to manufacturer to pass on higher savings to consumer
Shelf Display – Strategy	All CFLs lumped in a CFL section of shelf	CFLs integrated with other types of lamps, grouped by application and marked "good", "better", "best" for energy efficiency
Shelf Display – Quantity and Location	Low levels at low levels	Increasing amounts; increasingly at eye level
Packaging	Plain white boxes	Informative packaging that explains benefits
Advertising	Local and regional by individual utilities and retailers	Joining national marketing efforts like ENERGY STAR and Change a Light Change the World campaign

Trend	1980s-1990s	2000s
Manufacturer Advertising	Almost none	Manufacturers participating in regional and national efforts
Retail Availability	Few available at retail. Direct from utilities.	Rising influence of big box DIY stores, e.g., Home Depot and Lowes and eventually in some grocery stores
Media	Little attention, not quite ready	Helping to spread the message, riding the wave

Early utility messages focused on how to use the technology, for instance how to make the large, odd- sized CFLs fit into home fixtures, and the energy savings possible, although as mentioned above, discussions of lumens per watt were meaningless to many members of the public.

In an effort to shape their programs more effectively, utilities early on asked focus group participants about marketing messages. Consumers in a 1992 San Diego Gas and Electric study expressed preferences for phrases such as "gives off the same amount of light," "saves 75% of lighting cost," "cost savings," "value for your money," "lasts 10 times as long," "longer bulb life," and "energy savings." [9] Long life, bulb coolness, and potential to minimize bulb changes in hard-to-reach places were all mentioned as benefits by consumers in this study. [9] Energy efficiency, brightness, and long-life were seen as the most appealing attributes of CFLs by members of this focus group who had been given six different CFL models to try. [9] In fact the top reason for liking or not liking a bulb was brightness (or lack thereof). [9]

Based on user group feedback, utilities began including more customer benefits in their messages. For example, Madison Gas and Electric adjusted its print radio and TV advertising from simple description of the technologies and dollar incentives to customer benefits such as personal and environmental savings in its 1990 to 1994 CFL campaigns. [11]

In an article on green marketing, eco consultant Jacquelyn Ottman said "the first rule of green marketing is the first rule of marketing: focus on customer benefits i.e., the primary reason why consumers buy certain products in the first place. Do this right, and you'll motivate consumers to switch brands or even pay a premium for your greener alternative. Next, keep in mind that for green marketing to work, it is important that customers: [41]

- Are aware of and concerned about the environmental issues your product addresses
- Feel that by using your product they will make a difference ("empowerment")

- Believe your claims …
- Feel your product will work as well as non-green alternatives …
- Can afford any premiums." [41]

She recommends from an organizational standpoint that environmental considerations be integrated into all aspects of new product development and marketing by encouraging cross-functional teams of marketers and engineers. She suggests that new stakeholders be enlisted besides suppliers and retailers, including educators, members of the community, regulators, and NGOs [41].

A 1999 study by Heschong Mahone for the California Energy Commission noted that "utility companies (not lamp manufacturers) have been providing the primary path for public education about CFL lighting, with brochures and demonstration centers, but their reach has been limited and their focus has been on energy savings rather than consumer attitudes. While energy efficiency may be a social good, it is not generally the prime criterion by which people choose their lighting. Lighting has many other aspects which are more prominent – pleasant ambiance, aesthetics, brightness, long-life, safety, security, ease of replacement, etc. CFL lighting manufacturers must sell these aspects as well. [20] They went on to recommend that the CEC provide public service advertising focused on associated, non-energy features of CFLs such as their long life and cool, safe operation. [20]

The NRDC, in their 1999 report, also recommended that utility energy efficiency programs use consumer education and marketing to shift consumer focus from product price to product value, including non- energy-efficient attributes like longevity and convenience. [23]

Calwell of Ecos Consulting reasoned that product payback might be a good marketing message as paybacks of one year or less are possible with CFLs. The value of the energy savings of a single bulb might never enter the picture, because it is difficult to detect one bulb's impact on a monthly utility bill, payback in terms of replacement light bulbs alone is a more obvious and sellable benefit. For example, if the average consumer determines that a CFL lasts 10 times as long as a regular light bulb, he or she would be inclined to buy it if it costs less than 10 regular light bulbs. [24]

In a national consumer survey of lighting choices including incandescents by Philips, longer life rated highest in the attributes consumers were seeking from their lighting purchases, exceeding their desire to purchase "a major brand," to get a "good value" for their money, or to purchase a "high quality" lamp. [23] In a 1997 telephone survey of 500 purchasers of LightWise bulbs, 71% said energy savings is why they chose a CFL over an incandescent and

57% said long life. [17] In a 1998 study of 521 Southern California Edison LightSaver program participants, 75% said they bought the bulb to save energy, while only 10% bought it to change light bulbs less, and only 8.3% said they purchased CFLs because of a rebate or discount. [18]

Product naming that emphasizes key benefits and de-emphasizes the fluorescent technology has been used in attempts to boost sales. It could be argued that Philips achieved greater early sales of CFLs than its competitors by naming its product something other than "fluorescent" (given the fear many consumers had of that technology) and co-branding it with a range of other efficient technologies in the hopes of shifting people away from the conventional incandescent lamp largely dominated by GE. Philips initially called its product EarthLight to highlight its environmental benefits then changed the name to Marathon to emphasize its long life. Heschong Mahone suggested in its 1999 report that the industry adopt a new industry-wide name for CFLs that would avoid the word "fluorescent" all together. [20] And the NRDC noted that many energy-efficient lighting manufacturers have gone out of their way to avoid the use of the word "fluorescent" on product packaging, using phrases like "electronic light bulb" or "energy saving bulb" instead. Recognizing potential confusion with this, NRDC recommended that manufacturers develop a common language for lighting. [23] Confusion over naming was noted in the barriers section above.

Noting another trend, while some early designs attempted to make CFLs look more like incandescent bulbs by hiding the tubes under plastic domes, the authors of a 1999 NRDC report advised that marketing strategies should celebrate the differences. "Most CFL marketing strategies to date have focused on the great similarity of CFLs to incandescent bulbs. This has probably been a mistake. If a product is so similar to what consumers already use and costs more, why should they bother switching? It may be time to use educational and marketing campaigns to celebrate the distinct advantages of energy efficient lighting, and the disadvantages of incandescent light bulbs. Incandescent bulbs have distinct disadvantages: high operating temperatures, short bulb life, and the inconvenience of more frequent bulb changes. There is also a style and design angle that can be explored. Incandescent bulbs create "hot spots" of light within fixtures and scalloping (alternating patterns of light and dark) on adjacent walls, while CFLs and especially linear fluorescents can provide a more diffuse, continuous light." [23]

However, marketing must also shape consumers' expectations about a new product so that they are not disappointed. Several studies have indicated consumer dissatisfaction with the amount or brightness of the light produced by CFLs. Consumers who were promised a 60-watt or 75 watt equivalent were put off by the unexpected warm-up times of CFLs and the more diffuse light produced. As a 1999 NRDC study noted, due to their engineering, CFLs would never match incandescent and halogen sources for highly directional, focused lighting applications, whenever an extremely small, intense light source is needed. [24]

"SHEDDING LIGHT ON LIFE CYCLE BENEFITS"
JACQUELYN OTTMAN OF J OTTMAN CONSULTING, INC.

In an article on green marketing, Jacquelyn Ottman of J Ottman Consulting, Inc., discussed an early attempt at CFL marketing: "Philips Lighting's first shot at marketing a stand-alone compact fluorescent light (CFL) bulb was as clumsy as the funny-looking bulb that didn't easily fit most lamps. To boot, the "EarthLight" name confused consumers (I personally thought it was some type of plant light), and at $15 each versus 75¢ for incandescents, Earth Light couldn't climb out of a green niche. The brilliantly relaunched 'Marathon' CFL solved several problems: the 'super long life' positioning and incandescent- looking shape appealed to the convenience-oriented mainstream, while the promise of saving $26 in energy costs over its lifetime lured thrifty consumers. With the ENERGY STAR label to add credibility, as well as new sensitivity to rising utility costs and electricity shortages, sales in 2001 were up 12% in a flat market."[41]

4.2. Transition from Utility to Retailer Marketing

As noted above, many utility efforts in the 1980s and early 1990s were DSM-driven programs focused on getting CFLs into the hands of consumers as expediently as possible, often through CFL giveaways, direct installation programs, and mail-order programs, all of which set up the utility as the direct provider to the consumer. In 1991 the majority of CFLs (43%) were being purchased directly from the utility. At that time, 38% of incandescents were purchased at supermarkets. [8] A 1999 NRDC report noted that utilities driven by annual sales goals were reluctant to stray from the "time-honored formula: pay rebates to entice people to buy products. However, these start-and-stop efforts rarely ratchet toward a long-term change in consumer preference. If anything, they only prove that everybody loves a sale, and that many people are willing to wait for the next one before buying again." The NRDC recommended that utility residential lighting programs shift resources from product incentives to marketing investments and that utilities, manufacturers, retailers, and the government work together to build a message of value instead of price in customers' minds. [23] In the mid 1990s utility programs did begin moving to CFL programs that actively involved retailers, with point of purchase rebates and coupons distributed by retailers. This enabled consumers to take a more active role in their own CFL acquisition, it enabled utilities to transition from a provider role to an advisor role, and it recognized the role retailers can play in influencing purchase decisions.

While retailers praised utility coupon programs, they also acknowledged the need for consumer education by the utilities and they requested more marketing by manufacturers. [8] In November 1996 and May 1997 telephone surveys as part of the Lightwise Program, 60% of retailers said they did not have enough information about CFLs to adequately sell the product. They felt unprepared to explain CFL terminology such as wattage conversions, power factor, and ballasts. □To increase their marketing capabilities, retailers asked for consumer education and point-of-purchase materials such as brochures, demonstration models, and wall displays, along with price incentives. [17]

The California Residential Lighting and Appliance Program (CRLAP) is an example of one program designed to address these needs. Begun in 1999, the multi-year effort involved several California utilities in a massive statewide effort to build retailer infrastructure for marketing CFLs, including professional training of sales representatives and regular visits by field staff to each store to display merchandise attractively, tabulate inventory levels, and maintain point-of-purchase displays together with cooperative advertising programs to leverage investments by manufacturers and retailers in product promotions. The program was a decisive shift from bill-stuffers and utility-driven advertising to in-store advertising materials and events, including point of purchase banners and shelf-mounted signage, frequent sidewalk sales, and heavily publicized torchiere turn-in events. [28]

The utility shift to market transformation type programs encouraged more active participation by all market players. Recommendations for successful programs and players, which emerged from several program evaluation studies, include the following:

- Design multi-year programs around the lighting season (Sept to April) not the calendar year. [23]

- Leverage retailer and manufacturer capabilities by crafting a more visible and attractive presence for efficient residential lighting products in retail outlets. [23]
- Educate retail sales staff through training and field visits by utility staff or manufacturer representatives. [23]
- Conduct in-store product demonstrations. [17]
- Target a few key market segments to increase product availability, improve quality, and reduce long-term purchase price: dimmable and subcompact CFLs, recessed downlights, decorative porch lights and linear fluorescent fixtures, and lighting controls. [23]
- Provide point of purchase displays and signage.
- Leverage marketing dollars by combining the efforts of manufacturers and utilities and joining regional and national marketing efforts like ENERGY STAR.
- Promote CFLs through several different mediums to reinforce customer familiarity and positive associations with product. [10]

Several recommendations were aimed at utilities in their advisor/educator role:

- Utilities should focus on information programs, which were praised as "the cornerstone on which all energy conservation efforts are built" [11]
- Utilities should conduct consumer education activities including utility bill stuffers, and CFL product demonstrations at local fairs, home shows, and energy shows. [17]
- Utilities should support public service advertising to educate consumers and bring efficiency forward as a lighting selection criterion. [20]
- Utilities should provide consumer education tools to retailers like brochures, posters, demonstrations, and wall displays. [17]
- Utilities should offer education programs like telephone hotlines and websites to consumers and training programs for builders, designers, and retailers. [11]

It was noted that, like retailers, buyers groups can also be an influencer on the availability of CFLs. Buyers groups were formed to help smaller retailers obtain a greater variety of products and improved pricing by entering into contracts with large manufacturers on behalf of several retailers. For example,

True-Value stores typically purchase lighting and other products from one buyers group. Because buyers groups influence many smaller retailers (in 1999 there were 750 TruServ stores in the Northwest alone), they can have a significant influence on the success of CFLs and in 1999 they were noted as a group that had not yet been the focus of utility CFL efforts. [24] Betterbulbsdirect.com was established by the Northwest Energy Efficiency Alliance in 2000 to help small and medium sized builders find and enter into effective volume purchase arrangements for ENERGY STAR CFLS.

One aspect of the transition in program formats was the switch from utilities giving their financial incentives directly to the consumers to giving the incentive to the manufacturers. [10,17,18,22] For example the CRLAP targeted "significant investments of incentive dollars to competitively allocated manufacturer buydowns, which maximized price leverage per dollar invested.

More importantly, they greatly amplified competition among manufacturers. Those who moved their initial allocations of incentive dollars through retail channels by the program deadline could obtain additional allocations from their less successful competitors. This gave upstart manufacturers a foot in the door when competing for retail distribution space with larger, well-established competitors." [28]

1999 California Residential Lighting and Appliance Program (CRLAP) (Excerpted from Calwell et al. 2002)

"In 1999, after years of individual efforts, California utilities embarked on a joint, multi-year effort to transform the retail landscape for ENERGY STAR® lighting. Working primarily through subcontractors Ecos Consulting and ICF Consulting, the California Residential Lighting and Appliance Program (CRLAP) sought to do more than simply bring down the price of CFLs. It undertook a broad, sustained initiative to build a conscious market preference for ENERGY STAR® labeled bulbs, torchieres, and hard-wired fixtures. Program features included

- Assignment of channel managers and unique marketing strategies to each key retail segment.

- A decisive shift from bill-stuffers and utility-driven advertising to in-store advertising materials and events, including point of purchase banners and shelf-mounted signage, frequent sidewalk sales, and heavily publicized torchiere turn-in events.
- A massive statewide effort to build retailer infrastructure for marketing the products, including professional training of sales representatives and regular visits by field staff to each store to display merchandise attractively, tabulate inventory levels, and maintain point-of-purchase displays.
- Cooperative advertising programs to leverage investments by manufacturers and retailers in product promotions.
- Significant investments of incentive dollars into competitively allocated manufacturer buydowns, which maximized price leverage per dollar invested. More importantly, they greatly amplified competition among manufacturers. Those who "moved" their initial allocations of incentive dollars through retail channels by the program deadline could obtain additional allocations from their less successful competitors. This gave upstart manufacturers a foot in the door when competing for retail distribution space with larger, well-established competitors. [28]

Collectively, these program elements helped accomplish something previous California utility programs had not: they built a truly competitive marketplace in efficient lighting. This occurred in two key ways. First, it meant that Lights of America, a dominant early participant in utility incentive programs for efficient residential lighting, began to face strong competition from similarly nimble manufacturers of affordable products. Second, it meant that energy efficient lighting became an important enough revenue source to Home Depot, Lowe's, Home Base, Orchard Supply, and other retailers to cause them to invest their own promotional resources (discounts, advertising, etc.) to grab market share from their competitors. Beyond increases in product sales, the CRLAP program achieved the following in 2000 (Latham 2001): [28]

- 837 sales staff in 179 participating national retail stores trained, with a 54 to 76% increase in proficiency test scores regarding ENERGY STAR® lighting [28]
- 1,409 self-standing displays placed in retail stores

- 13,567 pieces of point-of-purchase (POP) material placed in stores, include banners, hang tags, and shelf-mounted cars
- In-store product demonstrations conducted in 158 DIY, mass merchandiser, grocery and specialty hardware stores
- $600,000 of cooperative advertising and promotion across 10 different projects, leveraging equivalent investments by manufacturers and retailers
- In-person recruitment meetings with 18 different ENERGY STAR® lighting product manufacturers [28]

The results of this large-scale effort to reshape the retail landscape became quickly evident. In 1998, California's CFL market share was about 0.8% -- double the 0.4% share observed in the rest of the U.S. However, by 2000, the California share surged to nearly 1.2%, widening the gap with the rest of the U.S., where the share hovered between 0.4 and 0.6%." [28]

4.3. Shelf Displays

Philips, in a national consumer survey cited by a 1999 NRDC report, learned through market research that only 30% of light bulb buyers enter a store knowing exactly what they want. The remaining 70% make that decision at the store, which drove Philips to revamp its point-of-purchase displays entirely to segment its products by application and purpose rather than by technology or wattage. Osram Sylvania made a similar change in its displays even earlier, and General Electric's displays also became more application-focused. Now a customer can go to the store seeking a bulb for a reading lamp, for example, and be presented with good/better/best options that correlate to longer product life and greater energy efficiency. This strategy was expected to increase CFL sales compared to the older strategy of placing all CFLs in a single part of the store, regardless of their light output, shape, or application." [23]

Several studies found that shelf space allotted to CFLs increased over the course of the study, that more retailers moved CFLs from temporary dump bins to permanent shelf locations, an increasing percentage of CFLs were moved from the less desirable knee-level display shelves to the more desirable eye-level shelves, more retailers carried more models of CFLs, and more retailers carried CFLs year-round instead of just during promotions. [17,32]

4.4. Packaging

Through the 1980s, CFLs were still being sold in plain white boxes. In the early 1990s, major manufacturers began explicitly packaging CFLs to the residential consumers with full-color packaging. [24] Consumer research has identified packaging as an ideal medium for informing consumers about this technology, which is still unfamiliar to some. Several studies have specifically addressed the impact of various packaging messages. For example, studies found that the CFL packaging image showing 10 incandescents = 1 CFL was well received [8]. The NRDC noted that many energy- efficient lighting manufacturers avoid the use of the word "fluorescent" on product packaging, using phrases like "electronic light bulb" or "energy saving bulb" instead. This was confirmed in interviews PNNL conducted with lighting manufacturers for this report. Manufacturers also told us they use product packaging to highlight selling points like energy savings and long life. One manufacturer changed its packaging to focus on what consumers recognize easily, making it clear that CFLs are incandescent replacements. Their packaging refers to watt equivalents rather than lumens; for example, it clearly states that a 15-watt CFL is a replacement for a 60W incandescent, and it lasts 6 to 7 times as long. Manufacturers also use graphics, such as a drawing of a CFL next to an incandescent showing that a 15 watt CFL equals a 60-watt incandescent. One manufacturer said rather than explaining lumens, they talk about brightness on the packaging, saying "bright" was a word that resonates with consumers.

A Point of Purchase Display that Talks to You

Lights of America produced a point-of-purchase display as part of Southern California Edison's LightSaver Program which included an audio message activated by pushing a button that explained the cost and energy savings benefits of compact fluorescent bulbs.

"Do you know you can save up to $127 each year just by changing a light bulb? Lights of America in cooperation with Southern California Edison, is offering compact fluorescent bulbs at a special price. Now you can cut your lighting costs way up to 80%. Compact fluorescents produce three times the light of a regular bulb with less heat and last up to 16 times longer. They turn on instantly and produce high quality flicker free illumination. Look for the special red and white sticker on Lights of America products to receive this special price". [18]

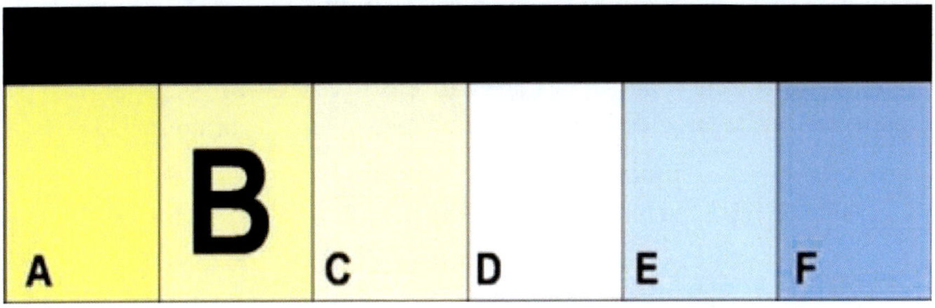

Figure 4.1. Proposed Lamp Color Label for CFL Product Packaging. Label uses a range of colors, from yellow (A) to blue (F), to indicate whether the product produces warmer or cooler light.

Manufacturers have also transitioned from boxes to blister packs, a move intended to reduce breakage in shipping that also has the advantages of providing more front-side surface for product information and allowing customers to see the product before they buy it.

A new move would add a color label to packaging to help identify whether the CFL product produces a warmer or cooler light. The effort, put forth by a color communication group consisting of representatives of lamp manufacturers, the National Electrical Manufacturers Association, DOE, EPA, and the Consortium for Energy Efficiency, would convey color through a label on the packaging that shows six colors, which are assigned letters. A large letter indicates the color of the light produced by that product (Figure 4.1). [45]

4.5. Advertising

Advertising efforts were marked by the trend from local utility-sponsored advertising to regional and national advertising that leveraged the efforts of all market players with the synergism of a national branding mechanism, the DOE-sponsored ENERGY STAR CFL program, and a national marketing campaign, the ENERGY STAR Change a Light, Change the World program. Seen in this trend was a move from limited to increased manufacturer advertising of CFLs.

As noted in Section 3, as early as 1989, utilities began joining forces in groups such as REAL, CEE, NEEP, the Northwest Energy Efficiency Alliance, WECC, MEEA, CRLAP, and the California Compact, to develop regional energy efficiency programs. "The relative cost of participating in regional and national collaborative forums is small," a 1994 Home Energy article summarized, "but the potential effectiveness of such efforts in simplifying and leveraging the response of trade allies is enormous."[11]

A Unique Medium for Getting the Message Across – Realtors

The Northwest Energy Efficiency Alliance worked with John L Scott Real Estate to do a New Home Owner promotion in spring 2001, where they gave real estate agents welcome bags containing ENERGY STAR CFLs and coupons to give to new home owners. John L Scott conducted a media outreach and advertising campaign and the program was featured in nine industry newsletters, regional journals, and local daily newspapers exposing 4.5 million consumers to the ENERGY STAR message. [29]

The lack of advertising by manufacturers was noted in the barriers section, but this changed as the technology matured and manufacturers became an active partner in CFL advertising. For example, the Northwest Energy Efficiency Alliance partnered with Osram Sylvania for an Earth Day 2003 promotion in the Portland and Seattle markets. The two organizations developed a full-page color advertisement that ran in *The Oregonian*, *The Seattle Times*, and *The Seattle Post Intelligencer*. In conjunction with the ad, the manufacturer offered a $3 mail-in rebate at participating retailers, which included both DIY and independent hardware stores. The Alliance's program combined $287,911 in funding from retailers and manufacturers with $202,197 from its own cooperative funds, for a total of $490,108, which was used to promote ENERGY STAR lighting products throughout the Pacific Northwest. [30]

Once ENERGY STAR instituted its CFL program in 1999, regional programs had a consistent standard with high brand recognition to rally around, bringing synergistic marketing efforts to a whole new level. In the fall of 2001, when the national media campaign Change a Light, Change the World was launched and CFL advertising achieved a nation-wide scope.

ADVERTISING + A QUALITY PRODUCT = LASTING DIVIDENDS

Central Maine Power showed that good first impressions and strong advertising reinforcement can pay lasting dividends. The utility instituted a coupon program called Operation Lightswitch in September 1994 with a major drugstore and two supermarket chains in its service territory. Central Maine let retailers get bids from several CFL manufacturers but insisted they require high quality electronically ballasted lamps. Although the CFLs normally retailed for $19-24 the retailers all agreed to lower their prices to $13 or $14. Then when one retailer started offering them at $3.99 with the $7 utility coupon they all switched to that price. Initially Central Maine mailed 450,000 coupons to customers in September and followed up with radio, tv, and newspaper advertising and point-of-purchase displays. When sales lagged in the first few weeks, a telephone survey revealed that 82% of customers had heard of the program but forgot to take the coupon with them when they went shopping. Coupons were then made available at all retailers and demonstrators were employed in high-volume stores. Sales skyrocketed with 171,000 sold by the program's end in December saving the utility 8 megawatts of capacity and 88 million kilowatt hours over the life of the bulb. Central Maine notes several retailers continued to stock CFLs and many have run their own promotions. [11]

4.6. The Growth of Home Improvement Stores and their Influence on CFL Sales

In several early studies, consumers complained about the unavailability of CFLs at grocery stores, where they normally bought bulbs. A 1998 Southern California Edison telephone survey of over 500 consumers found that the overwhelming majority were still purchasing their light bulbs at grocery stores (46.5%). [18] However, a report prepared by ECONorthwest for the Northwest Energy Efficiency Alliance published in August 2004 found that only 35% of consumers said they tended to purchase lighting at grocery stores. [35] This change may be due in part to the phenomenal growth of large home improvement centers. For example nearly 70% of the bulbs sold in the LightWise CFL bulb program conducted by the Northwest Energy Efficiency

Alliance in 1996-98 were sold at home improvement centers while less than 15% were sold at grocery stores. [17]

Utilities Use Point of Purchase Coupons and Multiple Promotion Avenues

Madison Gas and Electric was another utility who combined point of purchase coupons and heavy advertising. When it started its CFL program in 1990, less than 5% of its customer base had tried CFLs, only 4 out of 100 retailers sold CFLs and only 6 models were offered. Several modifications were made over the course of the 4-year program, including changes in technology offered (e.g., dropping higher efficiency incandescents which had been part of the pilot program), increases in incentives (from $5 to $10 per CFL), a change from mailed coupons to point-of purchase coupons, increased newspaper radio, tv, and newsletter advertising; increased contact with manufacturers and retailers; and the addition of retailer training. The changes paid off, resulting in 28% participation by customers, and 62 participating retailers selling 63 CFL models by 1994. [11]

How One Utility Used Change a Light

Connecticut Light and Power (CL&P) and United Illuminating (another Connecticut utility) used the Change a Light campaign to enhance their own programs. Both companies provided staff support during a Change a Light event at 17 Home Depot stores in Connecticut in 2001. More CFLs were sold during that eight-day period than CL&P had rebated or couponed during nine months in 2000. [30]

Home improvement centers, also known as do-it-yourself (DIY) stores, have emerged over the last two decades; the market leaders are Lowe's Home Improvement Centers and The Home Depot. With fiscal year 2004 sales of $36.5 billion, Lowe's Companies, Inc. is a Fortune® 50 company that serves approximately 11 million customers a week at more than 1,100 home improvement stores in 48 states. Based in Mooresville, North Carolina, the 59-year old company is the second-largest home improvement retailer in the world. The Home Depot was founded in 1978 in Atlanta, Georgia, and has since become the world's largest home improvement retailer and the second largest retailer in the United States, operating more than 1,900 stores such as The Home Depot, EXPO Design Center, The Home Depot Supply, and other subsidiary companies across North America. The Home Depot is the fastest growing retailer to reach $30 billion, $40 billion, $50 billion, $60 billion and $70 billion in sales. At the end of fiscal year 2004, revenue was $73.1 billion.

An NRDC 1999 study supports the finding that in recent years national chains, particularly home improvement centers, have grown tremendously in market share and influence. "The national chains now decide what products will be made available, and effectively marketed, to a large percentage of potential purchasers. Companies like Home Depot even have an on-staff energy manager, who helps keep the company's own energy bills low while promoting the benefits of energy efficiency to customers and retail staff as well. Significant opportunities to reach and influence national chains are trade shows and events such as the Dallas Market and the Chicago Housewares and Hardware Shows, where buyers for retailers are actively seeking to learn about new developments in lighting and select new products."[24]

The NRDC study notes that Home Depot makes many of its buying decisions through regional "team captains." These people share a dual responsibility within the company: to make all lighting buying decisions for the stores within their region, and to make recommendations to the other team captains on a particular category of lighting products within their area of expertise. As a result, conducting a successful program to promote ENERGY STAR CFLs within one region can increase the chances of such programs succeeding in other regions if the particular team captain involved is responsible for fluorescent lighting expertise within the company. Other retailers often entrust these decisions to a single national lighting buyer. [24]

4.7. Role of Media

No amount of marketing dollars can equal the impact of a front-page story in a major metropolitan daily or top billing on the evening news. The news media of course is a two-edged sword. In January 1999, *Consumer Reports Magazine* published a widely read and unflattering article that identified inconsistencies in CFL performance relative to claims made on product packaging. This article had unquestionable influence in prompting the discourse on the need for systematic testing of CFL performance. [24]

The media have also been a help in promoting CFLs. The trade and local press dutifully covered press conferences of the announcements of early CFL developments, even when they seemed to flare then fizzle, because the technology was premature[5,24]

However CFLs have had some very positive portrayals in the media which have very likely contributed to positive consumer recognition and sales. These instances can be attributed to more than good timing, they were also the result of having an improved technology, a solid groundwork of manufacturer-retailer utility relationships, a flexibility to shape programs quickly, and an understanding of what key marketing messages applied to the current situation. Calwell notes, "national outreach and education campaigns like Change a Light are likely to be successful if they can capitalize on existing market trends and waves of interest in particular products. It is easier to interest retailers and manufacturers in co-sponsoring promotions of products that are already highly popular than to rely on the national campaign alone to generate that interest. Monitoring sales trends and partner interests closely before launching such campaigns will obviously help point them toward promising technologies at the right time." [28]

Instances where CFLs capitalized on current events to gain positive media exposure included Earth Day 1990, the halogen torchiere debacle, and the Energy Crisis of 2001.

4.7.1. Earth Day

Earth Day 1990 represented something of a turning point in the national consciousness regarding energy efficiency in the home with messages of individual awareness and action to tackle global problems like ozone depletion and climate change. Curbside recycling became prevalent across the country. Books like *50 Simple Things You Can Do to Save the Earth*, *Consumer Guide to Home Energy Savings*, and *The Green Consumer* sold millions of copies and each of these recommended that consumers purchase CFLs. [24]

Subsequent Earth Day celebrations were additional rallying points for CFL promotions. For example in 1999 the Earth Day Campaign organized a promotion touting CFLs as a fundraiser and sent information to 20,000 schools across the United States inviting them to participate. In 2003, Osram Sylvania and the Northwest Energy Efficiency Alliance partnered for an Earth Day 2003 promotion in the Portland and Seattle markets. [30]

4.7.2. Halogen Torchieres

Appearing in the mid 1990s, halogen torchiere floor lamps were an apparent dream for consumers who snapped up the cheap, portable fixtures to light up apartments, dorms, home offices, and living rooms. The halogen lamps were however a nightmare for utilities, expending 300 to 500 watts per fixture. And at bulb temperatures of 750 to 1100°F, they soon became a nightmare for consumers as well; by 1999 they were reportedly the cause of at least 435 fires that killed 32 people, injured 114 more, and caused millions of dollars in property damage. Halogen torchieres were the subject of more than 200 safety- related lawsuits against halogen manufacturers and retailers, and prominent retailers like Home Depot and Wal-Mart dropped the product entirely. Television news programs devoted substantial coverage to safety problems with the products, and many universities banned their use in dormitories

Fortunately, CFL manufacturers were ready with ENERGY STAR-compliant CFL torchiere floor lamps that used a dedicated pin-based CFL to provide the same amount of dimmable light as 300 watt halogens, but at operating temperatures of only 100 to 200°F. The CFL torchieres were available without utility incentives for about $30 to $70, but thousands of customers across the country participated in highly publicized utility-sponsored torchiere turn-ins, turning in their old halogen torchieres for recycling and taking home an ENERGY STAR torchiere at sale prices. [24] In 2001 alone, the Northeast Energy Efficiency Project (NEEP) conducted over 60 torchiere turn-ins that resulted in the sale of more than 225,000 ENERGY STAR torchieres. [30] The CFL torchieres turned out to be an excellent means of introducing consumers to ENERGY STAR lighting; they were low-risk, highly affordable, and offered lifetime savings of $300 in avoided electricity and replacement bulb costs. [24]

4.7.3. West Coast Energy Crisis of 2001

The West Coast Energy situation began gaining media attention in 1999 when San Diego Gas & Electric became the first California utility to deregulate, allowing it to lift the price cap. Within a year, customers' bills tripled as the utility passed on high wholesale power costs. By mid June 2000, the crisis was making daily headlines with Californians experiencing rolling blackouts, utility bankruptcies, and stage three emergencies through 2001. The crisis spilled over into the Northwest as it tried to meet California needs while dealing with hydropower shortages from a summer drought and a cold winter. (http://www.pbs.org/wgbh/pages/frontline/shows/blackout/california/timeline.html)

Buying CFLs took on a patriotic flavor akin to buying WWII era liberty bonds as California's state and local governments mobilized a massive campaign to cut consumption and officials from the governor on down made public service announcements urging citizens to conserve. The California legislature mobilized $20 million in emergency funding for the California Conservation Corps (CCC) to purchase and distribute 1.9 million CFLs and 1.4 million "Flex Your Power" brochures to low-income households in every county in the state. [28] Conservation made headlines for helping to stave off blackouts (http://archives.cnn.com/2001/US/01/19/ power.woes.01/ index. html). News articles listed CFLs as an easy way to cut energy use and media web articles provided links to ENERGY STAR programs.

These events, and the media attention brought to them, combined with existing CFL programs that had helped drive increasing competition, brought down prices, sparking unprecedented sales of CFLs up and down the West Coast. By mid 2001 CFL market share reached 8% in California and 12% in the Northwest.

While no one can predict an energy crisis, there are lessons to be learned from this situation:

- Be mindful of current events.
- Be ready to share, have easy-to-understand consumer information and press kits prepared.
- Be flexible and able to adjust programs to meet current situations.
- Lay the groundwork by building relationships among market players early.

5.0. Where the Market Is Now

CFL sales have grown considerably in the last decade spurred by improvements in the technology and coincident changes in marketing. Some of these technology changes are highlighted in Table 5.1. Recent national market figures are given in Section 5.1. More specific data on market penetration from four American markets are provided in Section 5.2, and remaining barriers are described in Section 5.3. The growth in the CFL fixture market is described in Section 5.4.

Table 5.1 gives a capsule view of some of the major changes in CFL technology over the past decade.

Table 5.1. How CFLs Have Changed

Attribute	1992	2002	2005
Price Range	$15-$25	$3-$10	$0.98-$15
Life	9,000 to 12,000 hours	6,000 to 12,000 hours	6,000 to 12,000 hours
Type of ballast	Magnetic or electronic	electronic	electronic
Dimmable	no	few	some (significantly more expensive)
Incandescent Sized CFLs	none	some	many
Available wattages	9 to 34	2 to 59	2 to 59
Number of CFL manufacturers	Less than 10	hundreds	hundreds

Source: Platts with some data from Ecos Consulting June 2002 [30]

5.1. Market Share

Retail sales of screw-based CFLS were 4 to 6 million units in 1990, while retail sales of incandescent bulbs that year were 1 billion units, giving CFLs a retail market share of 0.1 to 0.2% [28] In 1998 CFL market share had grown to 0.4% and it hovered between 0.4% and 0.6% in 2000 before jumping to 2.1% by the fourth quarter of 2001 thanks to national marketing efforts, media attention from the West Coast Energy Crisis, and a glut of new, low-cost CFLs from foreign manufacturers. [28] There are no comparable data for CFLs' current market share, in part because key retailers Home Depot and Wal-Mart have removed themselves from the tracking surveys routinely conducted by industry analysts, but there is reason to believe that continued reductions in the average selling price of CFLs have pushed the products' market share steadily higher.

California's market moved more briskly, from 0.8% in 1998, to nearly 1.2% in 2000, then surging to 4% share in the first quarter of 2001 with a peak of 8.5% in the second quarter before stabilizing between 5% and 6% in the third and fourth quarters of 2001. Northwest sales rose higher reaching a market share close to 12% in 2001 before leveling off to at 5% to 8% of the market or 4 million sales annually after the energy crisis. [28]

National CFL penetration in 2002 showing data for number of CFLs in sockets, not sales, is shown in Figure 5.1.

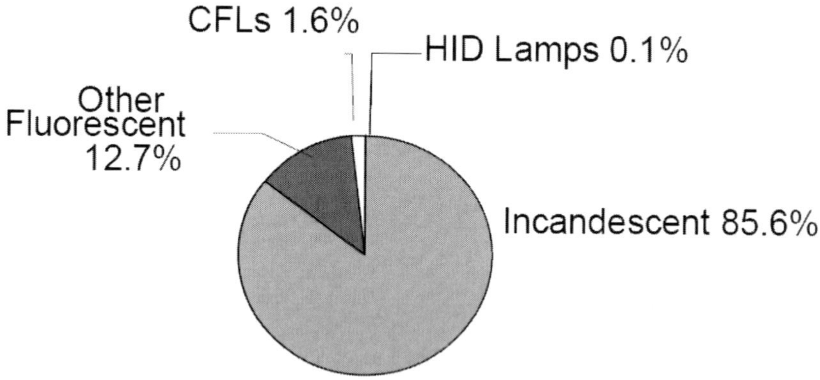

Figure 5.1. CFL Market Penetration. Incandescents filled 85% of sockets in 2002 compared to less than 2% for CFLs. Source Ecos Consulting [30].

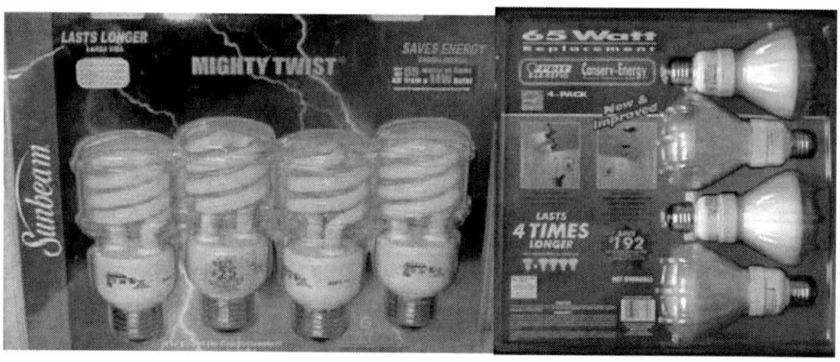

Figure 5.2. Can anyone in the U.S. get a CFL through a major retailer at a reasonable price with no utility support? Yes! Four CFLs for $9 at a K-Mart in Durango, Colorado, [27] and four Reflector CFLs for $11.99 at Costco in Portland, Oregon (PNNL photo).

Calwell summarized the state of the CFL market in 2002 (see Figure 5.2) "Stores like Home Depot, Ikea, Costco, and Kmart have made inexpensive CFLs a standard promotional item, in many cases bypassing major manufacturers to offer a private label brand at a lower price. Retailers now clearly understand the benefits of CFL technology, both to consumers (energy savings and convenience) and to the retailers themselves (higher sales revenues per unit of shelf space than incandescent bulbs). As a result, some level of in-store CFL promotion will now be self-sustaining, allowing the utilities to retarget or reduce their funding and still achieve a given level of market success. One overarching lesson from the events of 2001 is that market introduction takes time. CFLs ground their way to begrudging market acceptance for more than two decades before finally gaining traction with consumers. That period reflects multiple generations of revision to the technology itself, the trying and refining of numerous utility program approaches, and a great deal of hard work by a whole array of allies." [28]

5.2. Four Examples of Transforming Markets: The Pacific Northwest, British Columbia, New England, and California

The following four case studies give a more detailed look at changes in market share, sales, and consumer attitudes in four regional markets: the Pacific Northwest, British Columbia, New England, and California.

5.2.1. The Pacific Northwest Case Study

A survey of Northwest residential lighting customers conducted by ECONorthwest in June 2001 found that "a majority of the population (82%) is aware of CFLs and almost 30% of those who had purchased light bulbs in the three months prior to the survey chose to purchase a CFL. Sales in the Northwest r remained strong even after a high-visibility, $6/unit coupon campaign wound down. Most respondents surveyed indicated that they would continue to purchase additional products in the future, with or without coupons. More than 6.8 million CFLs were sold in the region in 2001 -- about ten times the estimated sales in 2000. [29] This led to a peak CFL market share of about 11.5% in the Northwest in 2001 -- perhaps the highest regional share achieved in the U.S. so far. [28] Market share leveled off to 5% to 8% of the market or 4 million sales annually after the energy crisis, according to a Northwest Energy Efficiency Alliance February 2004 program report. This is a drop from the 8 million sales during the crisis and coupon campaigns, but sales remained significantly higher than pre-program levels, and the Northwest continues to have the highest market share of CFL sales of any region in the country according to the EPA. [38]

A June 2002 market progress evaluation report by ECONorthwest for the Northwest Energy Efficiency Alliance's Residential Lighting Program reported that following the Spring 2001 West Coast energy crisis, regional sales were 20 times higher than Alliance preprogram planning estimates, totaling 8.6 million distributed (7 million sold plus 1.6 million given away) in the five quarters from fourth quarter 2000 through fourth quarter 2001. In a June 2001 survey of 1,421 residential customers, 179 had purchased CFLs within the past 3 months (30% of those who had purchased lighting of any kind in past 3 months). Only 3% of these were not satisfied with their purchase. Reasons they gave for satisfaction included longer lasting, energy efficiency, better light quality, and electricity bill savings. [29]

Reasons they gave for purchasing a CFL were to reduce their electricity bill (48%), in response to the energy crisis (25%), the savings were worth the higher price (19%), energy efficiency (10%), they wanted to try a CFL (10%), they felt it was the right thing to do (15%), and they wanted a longer lasting bulb (23%). (Respondents could give more than one reason). Only 12% said to redeem a coupon. [29]

Of those surveyed, 316 had purchased an incandescent in the last 3 months. When asked why they hadn't bought a CFL instead, 31% listed cost, 15% said they weren't aware of CFLs at the time of the purchase, and another 8% said they didn't have enough information. 11% said they couldn't find the right type, style, or size. Of those surveyed, 72% said they intended to purchase a CFL within the next year (that included those who had purchased CFLs recently and those who had purchased incandescents recently). 94% of recent CFL purchasers said they would buy another CFL even without a coupon and 64% said they would buy one in the future even without the energy crisis. Coupon sales accounted for only 39% of the 7 million CFLs sold during the five quarters. [29]

The drop in price for CFLs is one of the dramatic market shifts seen over the years, according to the Northwest Energy Efficiency Alliance. When regional efforts started, the price for a CFL ranged from $15-20 per bulb. In 2003 the average price for a CFL was around $5 a bulb or less than $3 a bulb if purchased in the now widespread multi-packs. [38]

For the fixtures market, in 1997 the Northwest Energy Efficiency Alliance noted that it started out with a handful of manufacturers producing 20 to 25 models, mostly torchieres. In 2003 ENERGY STAR listed over 5,000 qualified fixture models manufactured by several hundred different manufacturers. For the bulb market in 1997 there were 4 to 5 main manufacturers of CFLs, mostly those who participated in the PNNL sub-compact effort. By 2003 there were over 1,300 different models from more than 110 qualified ENERGY STAR bulb manufacturers, with at least 50 active manufacturers in the Northwest market. [38]

A report prepared by ECONorthwest for the Northwest Energy Efficiency Alliance published August 16, 2004, found that consumer awareness in the Northwest was very high as measured in several consumer surveys: 82% in May 2001, 88% in April 2002, and 88% in April 2003. The number of retailers stocking CFLs has increased, with 80% of retailers surveyed saying they planned to increase or maintain the number of models they carried. The price continued to drop with 48% of CFL purchasers saying they paid less than $6 per bulb compared to 20% in 2002. Market share had increased from close to 0% in 2000 to 9% in 2003. Regional sales went from 294,000 in the fourth quarter of 2000 to a high of 2,378,000 in the fourth quarter 2001, to a leveling off at 1,454,000 in the fourth quarter 2003. Consumer awareness of ENERGY STAR was high, consumer intentions to buy more CFLs in the future were high and retailers intended to offer the same or more CFLs models in the future. Consumer surveys done in 2003 showed 82% were more satisfied or as satisfied with their CFL purchases than with incandescent purchases. Reasons for satisfaction were: lasts longer 51%, energy efficiency 51%, better light quality 23%, saves on electric bill 18%, and works the same as an incandescent 13%. [35]

As the 2001 Energy Crisis hit, the Northwest Energy Efficiency Alliance program worked with more than 320 retailers. During the crisis this expanded to over 1400 retailers by March 2002, ranging from big box, DIY retailers, and independents, to general merchandise and groceries. Retailer participation leveled off at 1100 participants post-coupon campaign with a majority receiving field staff visits, the Alliance reported in February 2004. One of the best retailer market shifts can be seen with the participation of Costco. Costco did not sell CFLs until the advent of multipacks and lower per bulb prices. By 2004 Costco sales accounted for over 50% of the CFLs sold in the region. [38]

5.2.2. The British Columbia Case Study

A March 2004 report on BC Hydro's CFL initiatives in 2001-2004 examined availability, accessibility, and affordability of CFLs sold in British Columbia through an in-store field survey of residential lighting displays at 55 stores in BC Hydro's service territory in November 2003 and compared to a baseline study conducted in 2002. They found significant improvements:

- CFLs occupied 8.9% of total household lighting shelf space, up from 6.1% in September 2002, and 48% of this is at eye level.

- Comparable increases in relative market share were observed in the South Interior and Northern regions despite limited BC Hydro CFL programming in these regions.
- The majority of the increase in total shelf space for CFLs had been stocked with spiral CFL product. Spiral CFLS accounted for 4.9% of the shelfspace for all household lighting, up from 1.4% in Sept 2002. Spiral CFLs represented 54% of all CFLs for sale, up from 22% in 2002.
- Blister packs containing two, three and four CFLs accounted for an estimated 52% of all CFLs for sale.
- Over 170 different CFL models were available, an increase of 92% since 2002.
- Product priced at $6 or less accounted for 49% of all CFLs for sale. The majority of CFL product (84%) was priced below $10 per bulb.
- CFL purchase costs amortized over the life of the lamp had declined by as much as 76% since September 2002. [32]

A telephone survey of 802 BC Hydro residential customers was conducted in January 2004 and compared to previous telephone surveys conducted in 2002 and 2003 regarding household awareness of CFL technology and BC Hydro CFL programming.

- Awareness of CFL technology was at 90%, up from 83% in May 2003 and 74% in May 2002.
- The percentage of households that recalled information, advertising or promotions regarding CFLs from BC Hydro had increased from 22% to 65% since 2002.
- The percentage of homes using CFLs had increased from 23% in May 2002 to 55% in January 2004.
- Households using CFLs had installed an average of 5.2 CFLs each, an increase of 53% since May 2002.
- An estimated 4.2 million CFLs were installed in BC Hydro customer homes, up from 1.1 million in 2002. Sixty-four percent of all CFLs in use had been installed within the previous year.
- Thirty-seven percent of all BC Hydro residential customers had received a free bulb from BC Hydro.
- Nearly four out of every ten households (39%) who acquired a free CFL went on to purchase additional CFLs. While most purchasers in

the Lower Mainland acquired four additional CFLs each, a number of households purchased quantities of 15 and 20 CFLs each.

- As of mid-January 2004, Lower Mainland residents had installed 68% of all CFLs acquired through BC Hydro initiatives (giveaways or BC Hydro coupons). [32]

5.2.3. The Connecticut and Massachusetts Case Study

Recent data comes from New England where the percent of homes with at least one CFL now stands at 61% in Massachusetts and 63% Connecticut, according to a utility-sponsored in-home-visit survey reported in 2005. Sales climbed in Massachusetts from 158,000 in 1998 to 464,000 in 2002, then soared to 1.3 million in 2003, and to 3.1 million in 2004. Researchers attribute the shift to evolving efforts that transitioned from individual utility programs in the late 1980s, to a more concerted effort in the mid '90s with utilities making programs more consistent with each other and employing shared advertising, the same qualifying product lists, and the same rebate levels. In 1998 they officially joined forces through the Northeast Energy Efficiency Partnerships (NEEP) and collaborated with the national DOE ENERGY STAR program and associated marketing efforts. [42]

In the fall of 2002, the program underwent a shift toward industry-sponsored initiatives using an industry buydown process involving market actors. This facilitated a dramatic drop in dollars spent per unit moved - from $18 per product sold in 2000, to $13 per unit in 2001, $10 per in 2002, $4.84 per in 2003, and $4.61 per in 2004. Residential lighting programs now account for 22% of total utility energy savings achieved in New England. The mean number of CFLs per home is now 6.7 in Massachusetts and 6.8 in Connecticut, and 8% of all bulb sockets in Massachusetts and 7% of all sockets in Connecticut now contain CFLs. While these figures are impressive, the researchers noted some Massachusetts residents still cited barriers to CFL purchase, including more expensive up front cost 80%, limited selection 52%, don't fit 40%, aesthetically not pleasing 43%, and don't produce enough light 25%. [42]

5.2.4. The California Case Study

A 2004-2005 study by RLW Analytics for California's investor owned utilities gives a snapshot of recent California CFL usage. They conducted 850 onsite surveys of single family, multi-family, and modular homes in California. Among their main findings were that the overall number of all fixtures per home increased from 20 to 23 between 2000 and 2005 and the number of lamps (bulbs) increased from 34 to 41. On average, 11% of all fixtures had at least one CFL up from less than 1% in 2000, with most CFLs in floor and table lamps even though ceiling fixtures were the most common type of fixture. The 2004-05 study found 57% of all homes had one or more CFLs installed compared to the 2000 CLASS study which found 12% of all homes and the 2003 Residential Appliance Saturation Study which found 51% of all homes had a CFL installed. More than 90% of installed CFLs were in the 13- to 26-watt range and had screw-in bases and integrated ballasts. The most likely rooms to have CFLs were the living room (24%) and the master bedroom (20%). Areas with the highest reported lighting use were the kitchen, living room, outside, and garage. [37]

5.3. Remaining Barriers

With their increased retail availability and growth in sales in recent years, can we claim that a significant level of CFL market acceptance has been achieved? Most studies have said it's too early to declare victory. Barriers do remain. Reasons noted for not buying CFLs include cost, fit, didn't know about, and buying incandescents out of habit [29] and some consumers still indicate dissatisfaction with CFLs for various reasons: not bright enough, size, color consistency, and slow start-up time. [31,36]

The fact remains: if CFLs captured 2.1% of market share in 2001, then 97.9% of the U.S. bulb market was buying something else. Even in the Northwest, which has posted the highest quarterly CFL sales ratios in the country, 63% of households surveyed still had not purchased one CFL according to a 2002 study. [29]

While it is clear that progress has been made in this market, it is also clear that sustained, higher levels of CFL sales will be necessary to achieve market penetration levels that could be reasonably equated with transformation of the residential lighting market, according to a Northwest Energy Efficiency Alliance February 2004 program report. [38] The notion of socket share should be kept in mind here – every CFL sold displaces the sale of about 6 to 14 incandescents over its lifetime, so percentage of units sold can be deceptive.

One industry consultant noted recently "In our experience, we find that "market penetration" and other metrics can be misleading if not done on a consistent basis. As witnessed by the increased in sales (driven by the advent of "multi-packs" in many regions), one can conclude that the overall penetration by CFLs is quite high. However, this is only partially true, as it masks the clear division between product accessibility for consumers in urban and rural areas. Many rural areas, even in the Northwest, still have very low product availability and high prices that were seen in surveys of many years ago. We suggest that a consistent baseline with specific indicators be consistently carried out to serve as a true gauge of market transformation." (Personal. Comm, My Ton, 5/2/0/6.

5.4. A Growth Area - CFL Fixtures

Sales of dedicated fluorescent fixtures were more lackluster than sales of screw-based CFLs from the 1970s through the 1990s. From 1967 to 1997, fluorescent portable fixtures recorded essentially no gains in dollar sales, even as the overall market grew more than four-fold. When the impact of inflation is accounted for, the fluorescent portable fixture market shrank significantly. [24] Portable fixture sales picked up with the well-publicized torchiere turn-ins in the late 1990s; however, according to the Northwest Energy Efficiency Alliance, CFL fixture market share remained low, probably in the 1%-2% range and ENERGY STAR fixture sales were flat (at approx. 70,000-90,000 fixtures annually). Market barriers persisted in the areas of market fragmentation, performance, reliability, availability, lack of uniformity in pin-base configuration for lamp replacement, and lack of customer acceptance. In the early 2000s, DOE, EPA, and energy efficiency organizations such as CEE and the Northwest Energy Efficiency Alliance were pursuing programs to promote CFL fixtures in specific applications such as outdoor lighting and recessed can lighting. [38] Today, the CFL fixture market is getting a lot of attention thanks to efforts by these organizations in recessed can lighting; a new design competition; greater efforts at ballast, socket, and pin standardization; new mandated California Title 24 requirements; and a wide range of new offerings from CFL fixture manufacturers. [33]

5.4.1. New California Title 24 Requirements

The slowest sector in CFL products, pin-based CFL fixtures, is expected to lag no longer since California's Title 24 Residential Lighting Building Code took effect October 1, 2005. The new code (www.energy.ca.gov/ 2005_ standards/rulemaking/documents/index.html) requires all hard wired (i.e., permanently installed, not portable) fixtures in every room of the house over 70 sq. ft., except the kitchen, to be high-efficacy, or controlled by a motion sensor. In the kitchen, at least 50% of total installed wattage must be high efficacy and when calculating total installed wattage, the builder must assume for any incandescent or screw-based fixtures that the maximum rated wattage listed on the fixture by the manufacturer is used, not the wattage of the lamp put in the fixture at installation. For example, if a builder uses recessed cans with incandescent screw-based lamps in the kitchen, their total wattage must be less than half the total lighting wattage installed in that room.

The revised requirements apply to new construction single and multi-family homes, and hotel and motel guest rooms. Screw-based CFLs do not count as high-efficacy under the new requirements because they are too easy to replace with incandescent lamps.

Noah Horowitz of the NRDC, in an April 2004 presentation, predicted that this would greatly increase the demand for pin-based fixtures, especially recessed cans.

He recommended that manufacturers create marketing materials targeted to the California market ("meets Title 24") and cluster qualifying products in the ir catalogs. He also warned that compatibility is likely to be an issue again, because there are many proprietary pin-socket combinations. Look-alike products could also be problematic if one product complies with ENERGY STAR and a similar looking product does not. [33]

The new Title 24 requirements are summarized in Table 5.2.

5.4.2. Reflector CFLs

As CFLs became more popular after the 2001 energy crisis, manufacturers began offering more varieties of specialty CFLs for use in specific fixtures that were not very accepting of standard CFLs for a variety of reasons. Reflector CFLs (R-CFLs) were one key application since newer homes often had upwards of 25 recessed can fixtures. With the huge increase in the popularity of recessed can fixtures installed in US homes, and the lack of CFL fixtures or screw based CFLs available, DOE sponsored two energy-efficient reflector lamp programs: one to develop new insulated can lights requiring dedicated pin-based CFL lamps and another encouraging the development of higher performing screw-based reflector CFL lamps for recessed can very wide flood light applications.

Table 5.2. New 2005 California Title 24 Requirements Compared to 2001 Requirements [33]

Requirements	2001	2005	Comments
Recessed "can" lights	Must be approved for zero clearance insulation cover (IC rated)	Must be approved for zero clearance insulation cover (IC rated) and certified air tight (AT)	
Definition of high efficacy	Not less than 40 lumens per watt Must not contain medium base incandescent sockets	<15 watt fixture: min. 40 lumens per watt 15-40 watt fixture: min 50 lumens per watt >40 watt fixture: min 60 lumens per watt Same; i.e., no screw-based fixtures qualify	Low-efficacy = All screw-based lighting without hardwired ballasts and = All low-voltage incandescent, tungsten halogen, MR-11 and MR-16 High efficacy = all pin-based fluorescent
Ballast Requirement	none	Ballast for lamps of 13 watts or greater must be electronic, output frequency ≥20kHz	
Switch requirement	High-efficacy lighting must be operated on separate switch from any incandescent or low-efficacy lighting	Same	
Kitchen Lighting	General lighting must be high efficacy. Decorative lighting does not need to be high efficacy.	>50% of all installed wattage must be high efficacy. Must assume max wattage on any incandescent/screw-based fixtures.	Any incandescent or screw-based lamps must equal less than half the room's installed wattage.

Table 5.2. (Continued)

Requirements	2001	2005	Comments
Bathroom definition	Each room containing a shower or tub	A room containing a shower, tub, toilet or a sink that is used for personal hygiene.	
Bathroom	Bathrooms addressed separately from garages, laundry rooms and utility rooms	All hardwired lighting must be high efficacy, or controlled by a manual-on, auto-off occupant sensor,. Lighting in garages, laundry rooms and utility rooms must meet the same requirements as bathrooms	
Other rooms >70 sq.ft (hallways, stairs, dining, bedrms, closets	No requirement	All hardwired lighting must be high efficacy, or controlled by a manual-on, auto-off occupant sensor, or controlled by a dimmer.	
Outdoor lighting attached to a building	No requirement	Must be high efficacy, or controlled by both a motion sensor and photocontrol.	Lighting not attached to a building is exempt from this requirement.
Res. parking lots & garages for 8 + cars	No requirements	Must meet lighting requirements for non residential buildings	
Common areas - low-rise res. w/ 4+ units	No requirements	All hardwired lighting must be high-efficacy or controlled by an occupancy sensor.	

The recessed can downlight program requested fixture manufacturers to submit bids for ICAT downlight fixtures that would perform well in this high heat application. When it became clear that there were few choices in ICAT fixtures, DOE offered design assistance workshops to fixture manufacturers. Since then, the 2005 upgrade In California's Title 24 has led to numerous new ICAT cans; it is unclear how well these fixtures will perform in the next few years. The R-CFL Technology Procurement sponsored by DOE resulted in the introduction of R-CFLs designed specifically for operation in high-heat applications (ICAT cans). This project helped address a need for innovation in a niche category of CFLs where available products in early 2000 suffered from performance problems including fit and glare (protrusion below the recessed can opening) and from premature failure caused by excessive heat. Three lamp models, two from Philips and one from Feit, became available in 2005.

5.4.3. Lighting for Tomorrow

In 2002 the American Lighting Association, the Consortium for Energy Efficiency, and DOE (represented by PNNL) teamed to conduct a national lighting fixture design competition called "Lighting for Tomorrow" whose aim was to get more high-end, dedicated, energy efficient fixtures into the market. The lack of fixture options was identified through talks with lighting showrooms as a primary barrier to more energy-efficient lighting options for homes. Lighting showrooms are a principal distribution channel for the residential new construction market, which is estimated to be approximately 1.3 million housing starts annually (U.S. Dept of Commerce, U.S. Census 2001 Housing Starts). The first year of the competition welcomed both new and existing designs, as well as student designs with a manufacturer partner. The paper designs were due March 15, 2003, and winners were announced at the American Lighting Association Annual Conference in Tucson in May 2004. Entries for the 2005 competition were judged in June 2005 and were announced at the American Lighting Association Annual Conference in Miami in September 2005. (http://www.lightingfortomorrow.com) [30] Prizes were awarded to ten fixture families in the indoor and outdoor fixture family and technical innovation categories. Entries are being accepted through June 15 for residential decorative lighting fixture families for the Lighting for Tomorrow 2006 competition. Winning entries will be highlighted in a "yearbook" catalog delivered to lighting showrooms and retailers across the country. An LED/solid state lighting competition is also underway in 2006.

6.0. IMPLICATIONS FOR LED LIGHTING

6.1. Recent Feedback from Lamp Manufacturers

In fall and winter 2005-2006, PNNL interviewed product and marketing managers in the lighting divisions at Philips, GE, Osram Sylvania, and TCP (in most cases two staff at each company) to collect their thoughts on lessons learned from three decades of developing and marketing CFLS. They provided us with their thoughts on a variety of issues and challenges related to the market acceptance and increased adoption of CFL products discussed in this paper. They were also asked to extrapolate their experiences to new lighting technologies, such as LEDs (light emitting diodes).

While manufacturers did provide some insights, they stressed that LEDs are still very much a niche product, with applications limited to specialty uses such as automotive, electronics, signage, and holiday/decorative lighting, with no applications currently suitable for general illumination. In fact, they noted it may be 10 years or more before LEDs are ready for wider applications. One manufacturer stated "The buyers for these types of products are not the same as the buyers of CFLs. Buyers who might be interested in LEDs are more sophisticated and know their operating costs and the value of energy savings over time. In other words, they are most likely to be commercial rather than residential buyers."

6.1.1. Educating the Consumer

The manufacturers' representatives were asked about the need to educate consumers on the benefits of CFLs, and lessons for marketing other new lighting technologies.

CFL lighting manufacturers felt that many consumers are unaware of or unconvinced of the benefits of CFLs. Manufacturers we spoke to said they were faced with fairly limited advertising budgets for CFLs, so large-scale consumer education campaigns were not an option. Also, one manufacturer said that TV advertising didn't seem to work when it came to building consumer awareness of CFLs. They felt it was much more successful to work with key retailers and to do in-store demonstrations. Another manufacturer took a more sweeping approach, saying they use packaging, signage, promotions, radio and TV advertising. This manufacturer noted that they participated in a "Change a Light" TV campaign last year.

Lighting manufacturers often use CFL packaging to convey product benefits and to help the consumer determine the right CFL for their application. Selling points are clearly listed on the product packaging. Two manufacturers said they specifically focus on energy savings and long life. One manufacturer changed its packaging to focus on what consumers recognize easily, making it clear that CFLs are incandescent replacements. Their packaging clearly states that a 15W CFL is a replacement for a 65W incandescent, and it lasts 6 to 7 times as long. Consumers don't understand the value of 6,000- to 7,000 hour lifetimes until it is translated into years, according to the manufacturers interviewed. One manufacturer said they believe that it is critical to educate the consumer by providing literature at the retail level. This manufacturer has developed a CFL pamphlet that they affix near the store shelf so consumers can read about CFLs and the manufacturer's product benefits. One manufacturer stated that you have to educate the consumers so they will look for the product and know where to get it. If a buyer goes to a Home Depot or other big box store, he/she may not be lucky enough to find in-store help that is educated. Another manufacturer noted that pre-educating the consumer was half the marketing battle, because, once in the store, consumers take only seven seconds to make their purchasing decision.

Manufacturers said they would take the same approach when marketing a new lighting technology like LEDs as they have with CFLs – focus on the product benefits that consumers value most and prominently display these on packaging. Lighting manufacturers know that consumers can be hard to convince unless there is a compelling reason for them to switch from an incandescent to a new type of lighting product. They also believe that there is currently no compelling reason for a consumer to switch from incandescent to LED lighting. For example if a consumer wanted a blue-colored light, why would they pay significantly more for a blue LED bulb that is expensive when they can buy a blue incandescent at a much lower price? (White light in an LED is not even an option yet – and probably won't be for several years.) Another said as prices drop, consumers will become more interested in LEDs.

6.1.2. Educating Retailers

When asked about the importance of educating retailers to the success of a new lighting technology, manufacturers agreed that retailer education will be important for new products, as it has been for CFLs. Challenges are that educating retailers is expensive and there is a lot of variation across retailers when it comes to their understanding of lighting products. As one manufacturer noted, "We need to keep in mind though that we have been pushing CFLs since the 70s and most retailers still don't understand them."

Manufacturers agreed that, while it is important that both consumers and retailers understand the benefits of a new lighting technology, it is generally easier and less expensive to train retailers. According to one manufacturer, "retailers are key," both in terms of making the product available to consumers and in terms of providing space for in-store displays and shelf space for informative packaging. Another manufacturer noted that if the retailers are knowledgeable, they can sell the product.

Getting retailers to carry the product can be a challenge. Retailers need to be convinced that a product is going to perform as stated by the manufacturer. One manufacturer said "If a retailer doesn't like the product, or doesn't think it will sell, he won't stock it." If a retailer does sell a poor quality product, the retailer will suffer the consequences of product returns and no repeat sales. Manufacturers representatives recalled the glut of low-quality CFLs that hit the market in 2001 and noted that retailers need to be informed on what is and isn't a quality product, especially with new and unfamiliar technologies.

The package is a great place to educate, although retailers have asked about discontinuing the blister pack to gain more shelf space since this packaging requires more room. Lighting manufacturers are concerned that this would leave less space for advertising copy. It is also critical for the consumer to see the product before purchasing.

According to one manufacturer, if you can get a big, well known retailer like a Home Depot to carry a product, that goes a long way with consumers since there is the perception that if a certain retailer carries a product, then "it must be good." In one manufacturer's experience, the best venue for CFLs from a consumer education standpoint has been small, local hardware stores where experienced sales staff are more likely to be available. However, it was noted that the product should not only be introduced via this venue, and a first approach would be to target mass retail outlets.

6.1.3. Lumens versus Watts

Lighting manufacturers were asked how they addressed consumer confusion over lumens versus watts. As one noted, you can't change 100 years of consumer experience with incandescents, so rather than educate consumers about lumens, manufacturers are using packaging that refers to watt equivalents to help consumers determine which lamps are best for a selected application. Manufacturers also use graphics, such as a drawing of a CFL next to an incandescent showing that a 15-watt CFL equals a 60-watt incandescent. One manufacturer said rather than explaining lumens, they talk about brightness on the packaging, saying “bright” was a word that resonates with consumers.

When marketing new lighting technologies such as LEDs to consumers, lighting manufacturers agreed that they will need to compare performance to products that consumers are already familiar with, such as fluorescents. They noted it will be important not to overstate benefits or make exaggerated claims. With early CFLs, some manufacturers mislabeled products and overstated the equivalency (for example saying a 13 W CFL = 75 W incandescent). Exaggerated claims of product performance or poor application comparisons will ultimately result in a disappointed customer and less likelihood of a repeat sale. One manufacturer felt that CFLs may help to pave the way for LEDs by encouraging consumers to think of lighting as a technology purchase, which, like computers and TVs, should be replaced with new technologies as the old becomes less effective.

6.1.4. Negative Perceptions of Fluorescent Lighting

Studies have shown that the word "fluorescent" has negative connotations for many consumers and we asked manufacturers how they deal with this negative association in their marketing efforts.

Two of the manufacturers told us that they avoid use of the word “fluorescent” on packaging; instead they use terms such as “energy saving bulb” or similar wording. The focus is on trying to show the product’s benefits (long life and energy savings) and equivalency to incandescent lighting in terms of performance and color. One person told us they don’t use the term fluorescent specifically because of the poor connotations, even though their fluorescent products have overcome most of the problems consumers associated with fluorescents. Another manufacturer said they don’t mention fluorescent or CFL on their packaging but focus on wattage replacement instead. On the other hand, another manufacturer works to educate consumers on the benefits of CFLs and fluorescent lighting. According to this manufacturer, “in a number of markets where penetration rates for CFLs are high, attitudes are changing and the old negative connotations are going away. CFLs of today are a totally different product than they were early on.”

With regard to LEDs, manufacturers thought it was too early to tell whether consumers generally had a positive or negative image of the lighting category. According to one lighting staff person, “Consumers likely have few preconceived notions about LEDs. Right now, it’s probably indifference. These products are likely seen as “futuristic.” Another manufacturer felt that LEDs were still too unknown for consumers to have formed an opinion of them. This manufacturer said that manufacturers will probably need to come up with a new term to market them since the public won't know what a light emitting diode is; however, they speculated that the lighting industry was unlikely to collaborate on terminology. Once LEDs are ready for general illumination, long life and energy savings will likely still be important. According to one manufacturer, it's too early to tell for general lighting, but for holiday/decorative lights, the benefits of LEDs are nice color, long life, and low heat. One manufacturer felt that manufacturers should come up with a category of lamps term for solid state lighting, one that all of the industry could agree on.

6.1.5. Energy-Efficiency Programs

When asked about CFL market introduction programs conducted by utilities and other organizations, manufacturers agreed that energy-efficiency program support has been helpful to increasing consumer awareness and sales of CFLs. Manufacturers generally favored programs that focused on consumer and retailer education and training over direct incentives, such as manufacturer buy-down programs.

However if direct incentive programs are pursued, manufacturers said they prefer manufacturer buy-down programs over coupons because of lower administration costs. One manufacturer stressed that utilities need to ensure a level playing field so that interested manufacturers have the chance to participate. "There have been times when all the incentive money has been spoken for by the time the offer even hits the street." There also need to be controls for quality – incentives should not be paid on poor-performing products because this hurts repeat purchases and the category of lighting as a whole.

In regard to ENERGY STAR, one manufacturer felt that ENERGY STAR should focus on consumers' dissatisfaction with CFLs to keep junk out of the market, but instead ENERGY STAR focused on high end efficacy and performance specifications that might limit style choices. This manufacturer also felt that manufacturers have trouble meeting ENERGY STAR specifications and that only a narrow type of product, like twister lamps, can qualify while other products like three-way and dimmable lamps can't qualify. This manufacturer felt ENERGY STAR sets an incandescent equivalency that is too low; thus consumers are dissatisfied with the light level. For example a 60 watt incandescent delivers 850 lumens typically, but ENERGY STAR does not require this. (Authors' Note, we found 19 three-way lamps and 20 dimmable lamp designs in the ENERGY STAR database.)

One manufacturer felt that the three strongest factors in encouraging CFL introduction were giving them away, the West Coast electricity crisis, and California's Title 24.

6.1.6. The Issue of Price

Manufacturers cautioned that using coupons and manufacturer buy-downs to lower the purchase price can be a double-edged sword – they put the product into people's hands, but give a false idea of price, which does not encourage future purchases, at full market prices. Manufacturers felt that utility give-away and manufacturer buy-down program approaches pushed the product into the marketplace, but did nothing to encourage retailers to sell the product, especially when they circumvented normal retail channels. One manufacturer said that buy-down programs and coupons can do more harm than good; most manufacturers felt that public service announcements, in-store demonstrations, and consumer education are better approaches for utilities and energy-efficiency groups.

According to one representative, "We have participated in ENERGY STAR events in the past with some success in terms of creating consumer awareness. However there are problems when the promotions involve incentives that significantly lower price. Once the incentives are gone, consumers experience sticker shock when they face higher prices (without incentives)."

Another manufacturer said, "coupons, rebates, etc., may hurt you if they lower price too much and the consumer does not make a repeat purchase because the (retail) price is perceived to be too high. If a consumer pays $1 for a CFL once, they expect that to be the price next time." As stand-alone strategies, the manufacturers felt these tactics do not work well; they simply don't have the power to 'move the needle.' Coupons and similar promotions should only be used as part of a bigger, overall marketing strategy and should not be used as the sole motivator to move the market. According to most manufacturers, education and demonstration/training activities are more effective than buy-down programs, which just cut into profits.

Manufacturers cautioned that it is premature to promote LEDs or develop any efforts with efficiency groups since the product is still in its infancy. Noted one manufacturer, "We don't want to promote them until they are ready – and they aren't for general illumination."

Manufacturers noted that retailers must be convinced to carry products, and unless there is a profit involved for them – it's a hard sell. Pricing LEDs is going to be an issue since there needs to be a way for retailers (and manufacturers) to make a profit on them. If the LED is inexpensive and lasts 50,000-100,000 hours, it could lead to overall reduced lighting sales and profits for retailers. If retailers make less on LEDs than they do on incandescents, there is no strong motive for them to carry the LEDs.

My Ton of ECOS Consulting suggests that utilities can add value to the product by making it "very clear through marketing messages that the utility is bringing the consumer a technically sound, environmentally efficient, money-saving product that saves electricity, reduces the consumer's bill, and decreases the utilities' energy distribution load." My Ton felt that CFL coupons that decrease year over year do transform the market while encouraging purchase by those who have never before tried CFLs.

6.1.7. Poor Performance

We noted that consumers had a number of complaints about the performance of early CFLs, often resulting in long-term distrust of CFLs as a lighting category in general. Manufacturers were asked what might be done differently when introducing a new lighting technology to avoid the same fate suffered by the CFLs.

One representative noted "Of course the solution to this is for manufacturers and retailers to present good- quality products to consumers. Otherwise poor quality goods will prejudice the buyer, and he may not buy again (possible long-term ill-will)." The manufacturer went on to say that this isn't typically how things work. "Not surprisingly, manufacturers often want to jump into the market with a new technology to be one of the first entrants, which means they might introduce a technology prematurely, causing the consumer to develop negative perceptions of the entire lighting category if it does not perform as expected.

One manufacturer felt that reflector CFL lamps should not be included in ENERGY STAR because of poor performance. He noted that LEDs will make a much better light source for reflector lamps because of their directionality.

Another manufacturer said that LED quality guidelines are needed and that this was a lesson learned from CFLs.

According to another manufacturer, because light bulbs (CFLs, incandescents, etc.) are low interest, 'commodity' items, the marketer MUST know the key issues that will motivate a purchase, and this is only achieved by asking a buyer what his/her needs are. The manufacturer referred to this as "soliciting the voice of the customer" and viewed market research as critical to understanding buyer lighting needs. Their company believes that taking the time to do this can be a challenge since retailers sometimes pressure lighting manufacturers into releasing products prematurely. Lighting manufacturers agreed that it is first important to do market research to discover the market drivers for potential buyers, so that the key purchase motivators are identified. It's best to delay a product rollout until a value proposition has been developed.

According to one manufacturer, CFLs really didn't suffer in the long run. "Yes, in the early stages they did not perform well but that's the natural path of technology innovation – early products are expensive and may not perform very well. Manufacturers throw products out there to see how they do. You aren't going to have winners in the market early on until after there are some dogs that help fund more R&D and over time the technology improves as do products."

6.1.8. Potential Approaches for Introducing New Lighting Technology

We suggested several approached that might help introduce a new lighting technology, such as LEDs, without "harming" consumer perceptions of the new category of lighting. Below are the manufacturers' reactions to the various approaches.

1. A Consumer Education Program Led by Industry or Others

"A good first step once products are ready. However I am not sure how you would organize this with industry. Might be able to work something out through NEMA."

"The more people that are educated, the better it is to move the product into the marketplace, so this would be a good approach."

"Its hard to know at this point since its too early to educate consumers on LED products. The Government should focus on supporting R&D to help improve the overall technology."

"The more education programs the better. No one wants to waste time buying light bulbs. Few people take the time to read the packaging and understand the technology. They just want to get in and out of the store."

2. Delay Product Introduction Until Performance Improves

"This is key. Everyone wants to enter the market now, but that doesn't mean that just because you CAN launch something, you should. Premature release of a product that isn't going to satisfy consumers can hurt both the manufacturer and the industry."

"Very important. It will be interesting to see how this pans out and whether some companies try to oversell benefits, which could hurt the market overall."

"This is unrealistic. Some manufacturers will undoubtedly launch lousy products with inferior technology."

"Yes, but it must remembered that any product gets better over time. If you don't introduce the product at some point, you can't earn money to sink back into R&D. "

3. Offer Only Restricted Sales to the Consumer Sector Until Performance Improves

"Yes, again this is important but easier said than done since some manufacturers will sell wherever they can with poor products. Then other manufacturers follow suit. This is hard to stop."

"Yes, we are selling LEDs for limited applications to the commercial sector; however there is no way to restrict products being marketed to consumers by others."

"Yes, the consumer sector is much more demanding."

4. Promote the Product Only in Niche Market Applications Initially. "Yes. Colored Leds Are Already Being Sold"

"We are focusing on blue LEDs right now. Again, however, who knows what others will do." "This is happening now."

5. Intensify Retailer Training So That Salespeople Understand the Benefits/Drawbacks and Best Applications for the Technology

"Yes, we will be training retailers in 2006. We can only teach so much though, and we aren't the only ones out there training salespeople."

6. Partner with Utilities and Energy-Efficiency Groups to Promote the Product

"Yes. We will do this, but not yet – further down the road, when the products are closer to release date. Right now this is premature."

"Yes, eventually when LEDs are ready."

"Yes but with a caveat – only for educational purposes. If the group … understands that, from a business standpoint things take time, it could be beneficial. If they have unrealistic expectations, it's not a good idea."

Figure 6.1. An LED Reflector Lamp. This design takes advantage of the directionality of LEDs by using them as the light source for a reflector lamp.

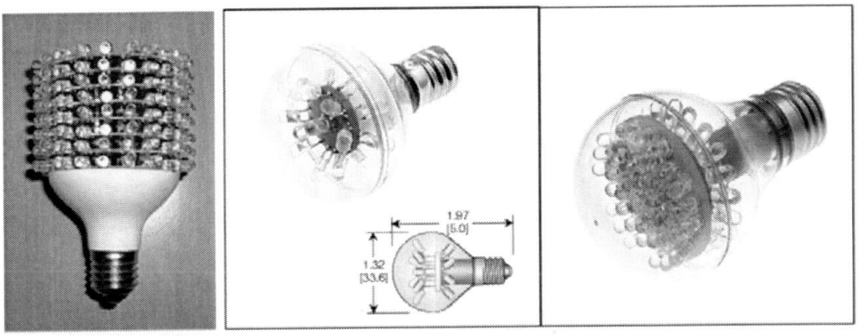

Figure 6.2. Two Examples of LEDs Configured as Replacements for a White Light Incandescent Bulb. While it is possible to configure LEDs in a shape that is similar in dimensions to an incandescent, this application does not take advantage of LED's unique qualities over incandescents, and comes at a cost that is currently much higher than an incandescent bulb.

6.2. Early Examples of LED Lighting

It appears that some early LED product offerings, in particular those that are being marketed as a replacement for incandescents for general service lighting, may be following a path similar to early CFL offerings and similar mistakes may be made. LED products for general service applications are very expensive, make overstated lumen equivalency and energy savings claims, and don't perform very well. The early applications of LED lights that have succeeded have capitalized on the unique nature of LED lighting - its high directionality, small size of the lamps, strong hues, brightness, solid state operation, and long life. In cases, where manufacturers are trying to force-fit them into general service applications, they are less likely to be successful and could in fact turn off consumers who might otherwise find LEDs perfectly acceptable for niche applications. Figure 6.1 shows an example of a reflector spotlight lamp, which takes advantage of LED's focused light source. Figure 6.2 shows some examples of LEDs that are being packaged as a replacement for an incandescent lamp.

7.0. Lessons Learned

Lessons learned from reviewing the prior efforts of the many involved in increasing consumer acceptance of residential CFL lighting, and interviews with CFL manufacturers, retailers, and others are summarized below. These lessons are grouped into three categories: those pertaining to technology, program design, and marketing.

7.1. Technology

Introduce new lighting technology in niche applications where benefits are clearly defined and performance is not overstated. Initial exposure is highly influential in the market and first impressions count for a lot. Often, there are niche applications where consumers will likely be quite satisfied with their purchase. For CFLs, this could have been porch light applications in moderate-temperature climates where performance drawbacks such as buzzing, flickering, color rendering, and fit may have been less of an issue. Early use of CFLs in table lamps may also have been acceptable. Use of early-generation CFLs for general illumination in high-use areas such as kitchens, and spotlight applications, would have commonly been met with consumer dissatisfaction. This early dissatisfaction could result in a consumer that is unwilling to try a CFL again – even once performance has been improved. (See 1.1.2, 2.1.)

Be aggressive about dealing with technology failures that affect main benefit claims. If products are failing prematurely or otherwise not performing as claimed and as expected by the consumer, changes in the product and its marketing should be made immediately before more buyers are "turned off." (See 1.1.2, 2.1, 3.1.1.)

Performance is more important than appearance. There was some debate with early CFLs as to whether consumers would want something that looked "just like an incandescent," or whether new designs like twister lamps might succeed. Some early focus groups found that consumers liked lamps with familiar shapes; however, in the long run the nonstandard (i.e., twisters) have sold due to their size and performance. It seems clear that consumers are willing to look beyond an odd appearance if a product performs well. But they are less forgiving of poor initial performance. (See 2.3.) In effect, the light bulbs themselves rarely have an aesthetic because they are hidden in fixtures, but the light delivered by the fixture itself does have an aesthetic, which is heavily influenced by the performance of the bulb.

Know and admit technology limitations, e.g., CFLs can't provide focused, highly directional, light the way halogen, incandescent, and LED sources can. (See 4.1.)

Manufacturers should work closely with energy efficiency groups to help establish minimum performance requirements that the manufacturers can meet and the efficiency groups are willing to support. (Lightwise changed requirements on power factor and THD after losing a major manufacturer participant and discovering that the added requirements were increasing product cost without producing a corresponding consumer benefit.) (See 3.2.)

7.2. Marketing and Consumer Education

Identify niche markets (building sub-markets) where the benefits of the new technology make especially good sense, and where initial sales can be built. The hotel and multi-family housing markets, where energy savings and reduced maintenance costs were important drivers to the use of CFLs, were initially targeted by DOE and some utility programs. (See 2.1, 3.1.1, 3.1.4.)

Focus message on areas where the technology can meet expectations, and reshape consumer expectations where new lighting technologies fall short. For example, CFLs are often a good choice for general illumination, but not for applications where a focused beam is needed (i.e., spotlight and narrow floodlights). (See 4.1.)

Marketing literature should show comparisons of savings to the standard technology, e.g. the 1 CFL = 10 incandescents. (See 4.1.)

Consumers find that long-life claims are hard to believe and hard to relate to, unless backed up by guarantees or trusted labeling. (See 2.3.)

Consumers want to see how a new technology performs before they buy it; (i.e., they want to see lit CFLs in in-store displays). (See 4.1.) It is generally also better to show the new technology clearly in transparent "blister packs" than to hide it in an opaque cardboard box. Don't apologize for differences from standard products; celebrate them.

Consistent, meaningful terminology across the industry is important for new products. Get consumer feedback on terminology early on and use education to counter any image problems. Consider adopting a new industry-wide name, for example with CFLs that avoids the word "fluorescent" all together. (See 2.2.2, 2.4.2, 2.5.1, 4.1.)

Product advertising should be specific about benefits. How much longer do they last? How much energy do they save? (See 2.5.1, 4.1.)

Shift consumer focus from product price to product value, including non-energy-efficient attributes like longevity and convenience. (See 4.1.)

Consider using educational and marketing campaigns to celebrate the distinct advantages of energy- efficient lighting and the disadvantages of incandescent light bulbs. (See 4.1.)

Conducting in-store product demonstrations is a good way to educate consumers. (See 4.2.) Promote CFLs through several different mediums to reinforce customer familiarity and positive associations with product. (See 3.1, 3.2, 3.6.3, 4.2.)

Consumer education activities might include utility bill stuffers, and CFL product demonstrations at local fairs, home shows, and energy shows. Education tools like brochures, posters, demonstrations, and wall displays should be provided to retailers. (See 2.4.1, 4.2.)

Public service advertising should be supported to educate consumers and bring efficiency forward as a lighting selection criterion. (See 2.4.2, 2.5.1, 4.1.)

Utilities should offer education programs like telephone hotlines and websites to consumers and training programs for builders, designers, and retailers. (See 4.3.)

Arrange shelf displays by lighting application rather than manufacturer and identify good/better/best options that correlate to longer product life and greater energy efficiency. For example, a customer going to the store seeking a bulb for a reading lamp would be presented with good/better/best options for reading lamp bulbs. (See 4.3.)

Ideally, new lighting technologies should be available where consumers typically purchase bulbs – in grocery stores. However grocery stores are unlikely candidates for new lighting technologies. Therefore, consumers will need to need to be educated on where to shop for new energy efficient lighting products (hardware and home improvement, mass market discounters where higher priced items are norm instead of grocery stores). (See 2.2.1, 2.3, 4.2, 4.5.)

Build a message of value (costs more/worth more) rather than the short-term fix of a rebate. (See 3.1, 4.1.)

Mass media is vital to widespread adoption of a new technology. Pay attention to both paid and unpaid. Paid advertising would include advertising on TV, radio, newspapers and magazines. Unpaid means getting messages to media through news releases, press events and being ready with media messages when spotlight turns to energy efficiency (like the California energy crisis of 2000 or Earth Day) or safety (like halogen torchieres). (See 4.6.)

Packaging needs to be more clear concerning warranties and replacement of failed products.

Use a color label on packaging that clearly indicates the warmth or coolness of a CFL product's light output. (See 4.4.)

7.3. Program Design

Study market structure to see how best to introduce a new technology. If it is the consumer market you are after, then understanding the retail markets and distribution practices is critical. (See 2.3.)

Utility giveaway programs can increase buyer awareness but don't create lasting market transformation by themselves, because they bypass normal market distribution routes. They may also undermine retail sales. (See 3.1.1.)

If providing incentives, utilities should consider developing a program that focuses on retailers and manufacturers. This allows for a lower price to the consumer and involves manufacturers in the process. Retailers are another good focus for incentives since the incentives can be coordinated with retailer training. (See 3.1.3.)

As with most new technologies, prices are highest initially when sales volumes are low and manufacturing processes may not be optimized. Manufacturers should keep in mind however that it's hard for consumers to assign a value to a product (i.e., know when they are getting a good deal) if price keeps fluctuating (from free to $25+). Utility rebates are likely to have exacerbated this problem with consumers. (See 2.2.1.)

Energy efficiency is often not the leading factor in energy-efficient lighting sales. Safety, for example, led to 1 million units sold in a year for the

1998 introduction of ENERGY STAR torchieres. Product marketing should focus on the product attributes deemed most important by consumers. (See 4.1, 4.6.2.)

Products shouldn't be given away unsolicited. Some action on the part of the consumer should be required, if not a purchase then mailing in a card. Action yields higher installation rates and greater awareness. (See 3.1.1.)

Utility program or manufacturer field representatives can be a very helpful link between manufacturers and retailers; they can educate and train retailers, set up displays, distribute promotions, and iron out problems. (See 3.1.4, 4.2.)

Utilities have found that developing enthusiastic and ongoing retailer participation is a key to market introduction programs. Lighting manufacturers, who generally have regional sales staff, should ensure that these staff are well trained on the new technology and can provide information to retailers. Utilities have found that building retailer trust can take 2 to 3 years, so ongoing communications are important. (See 3.1.2, 3.1.4.)

Programs that start with low-quality, poor performing (often low-cost) products to keep costs down may find that this approach can backfire. It is difficult to convince consumers to buy something they once received for free, especially if the initial experience wasn't a positive one. (See 2.1, 2.2.1, 3.1.1.)

Design multi-year programs around the lighting season (Sept to April) not the calendar year. Coordinating these programs with national campaigns, such as the ENERGY STAR Change a Light, Change the World program can leverage efforts for greater recognition and sales. (See 2.5.2, 3.6.3, 54.2.)

Lighting manufacturers, retailers, and utility energy efficiency programs should combine capabilities by crafting a more visible and attractive presence for efficient residential lighting products in retail outlets. Point-of-purchase displays and signage can be very helpful in driving initial sales. (See 3.2, 4.2, 4.3.)

Several studies showed that retailers were a primary source of information about lighting for consumers but surveys of retailers showed that they felt inadequately prepared to discuss energy efficient lighting. Retailers need and want more training so that they can feel more knowledgeable about energy efficient lighting. They also want tools to get the message across to the consumer: in-store displays, signage, brochures, more informative packaging, etc. (See 2.5.1, 4.2)

The enthusiastic and on-going participation of retailers was recognized by energy efficiency groups and utilities as one of the building blocks of successful programs, and one that needs to be acknowledged sincerely and frequently. [See 3.1.2.)

Consistency is important among overlapping incentive programs. Multiple programs with conflicting incentives can confuse and frustrate customers and make program delivery more complicated. (See 3.1.4, 3.6.4.)

Once regional efforts are in place, a new lighting technology with strong energy-saving potential is likely to benefit from regional support and coordinated approaches to promotional programs, versus a variety of approaches introduced by individual utilities. Lighting manufacturers should join forces with utilities and national energy efficiency programs (e.g., ENERGY STAR) for best results. (See 3.6.3, 3.6.4, 4.2, 4.4.)

Manufacturers are concerned about how to price LEDs so that retailers will get return on shelf space profit.

7.4. Application to LEDs and Other New Lighting Technologies

The following lessons learned are culled frominterviews PNNL conducted with product and marketing managers in the lighting divisions at Philips, GE, Osram Sylvania, and TCP in fall and winter 2005-2006 based on their experiences developing and marketing CFLS. Although they all commented that LEDs are a developing technology currently restricted to niche applications, the insights they provided could be valuable to the lighting industry in seeking market acceptance of LEDs or other new lighting technologies.

A lot of consumer research is needed to determine what the consumer does and does not know before the initial product launch so that the launch is done right the first time.

Accurate incandescent equivalency on packaging is critical.

Rely as much as possible on retailers for customer education.

Industry collaboration, perhaps through NEMA, would be helpful, though difficult to achieve given the large number of manufacturers.

Coordinate with energy-efficiency programs once products are available but don't start before products are ready.

Don't rely on giveaways and coupons or other programs that confuse consumers about the actual retail price. If the consumer's first experience of the product is through a give-away or drastically reduced price coupon offer, they will balk at paying a high retail price next time.

Avoid market introduction programs that distribute products outside normal retail channels, for example utility mail-order programs.

Performance claims must be accurate. Don't launch a product just because you can. Wait until performance issues are ironed out.

Initial education and performance issues will be more difficult to iron out if many manufacturers are involved in the initial introduction of LEDs.

Pricing is critical but tricky. For the manufacturer to make a profit on LEDs it must be priced low enough to encourage consumer demand but the retailer must be able to make money on it as well.

Education, of both consumers and retailers, is critical.

Understand that many people will not try a new product until price drops to a range near that of existing products providing similar functionality.

Niche marketing is the best approach for now.

REFERENCES

[1] Smithsonian National Museum of American History. (2003). "Competition for a Modern Lamp." *Lighting A Revolution*. Available URL: http://americanhistory.si.edu/lighting/20thcent/comp20.htm (last updated March 2003).

[2] Harris, Tom. (2005). "How Fluorescent Lamps Work." *HowStuffWorks*. Available URL: http://home.howstuffworks.com/ fluorescent-lamp.htm, searched June 2005.

[3] "Fluorescent Lamp." (2005). *Answers.com*. Available URL: http://www. answers.com/fluorescent%20lighting, searched June 2005

[4] Smithsonian National Museum of American History. (2003). "Inventing Six Modern Electric Lamps." *Lighting A Revolution*. Available URL: http://americanhistory.si.edu/lighting/20thcent/invent20.htm (last updated March 2003).

[5] Beardsley, C. (1976). "And Now, the Hollister Lamp." *Lighting Design & Application* April, 48-49.

[6] Bouwknegt, A. (1982). "Compact Fluorescent Lamps." *Journal of IES* July, 204-212.

[7] Goett A, Van Liere, K. & Quigley, D. (1992). "Customer Acceptance and Use of Compact Fluorescents: Results from a Comprehensive Evaluation of PG&E's Program." *Program Evaluation - Proceedings from the ACEEE Summer Study on Energy Efficiency in Buildings*, Volume 7, pp. 7.61-7.74. American Council for an Energy Efficient Economy, Washington, D.C.

[8] EPRI. (1992). *Perceptions of Compact Fluorescent Lamps in the Residential Market*. EPRI TR 100734, prepared by Macro Consulting, Inc., Mountain View, California, for the Electric Power Research Institute, Palo Alto, California.

[9] SDGE. (1993). *Appliance Efficiency Incentives: Residential Compact Fluorescents. 1992 Lighting Focus Groups and Follow-Up Interview Report*. MIAP-92-P09-S01-R241, prepared by Marketing Information Masters, Inc. for San Diego Gas & Electric, San Diego, California.

[10] Rasky, L. (1993). "Consumers and Compact Fluorescents." *Home Energy Magazine Online* November/December 1993. Available URL: http://homeenergy.org/archive/hem.dis.anl.gov/eehem/93/931106.html

[11] Fernstrom, G. B. (1994). "Steps to Successful Lighting Programs." *Home Energy Magazine Online* November/December 1994. Available URL: http://homeenergy.org/archive/hem.dis.anl.gov/eehem/94/941110. html.

[12] Atkinson, B. (1994). "Downsizing the CFL." *Home Energy Magazine Online* January/February 1994. Available URL: http://homeenergy.org/ archive/hem.dis.anl.gov/eehem/94/940103.html.

[13] Goett, A. A. (1994). "Seeing the Light: Effective Strategies for Promoting Compact Fluorescent Lighting to Residential Customers." *Proceedings from the ACEEE Summer Study on Energy Efficiency in Buildings*, Volume *10*, 10.41-10.47. American Council for an Energy Efficient Economy, Washington, D.C.

[14] Polsby, E. (1994). "Putting Energy-Efficient Lighting in Its Place." *Home Energy Magazine Online* November/December 1994. Available URL: http://homeenergy.org/archive/hem.dis.anl.gov/ eehem/94/ 941112.html.

[15] Haddad, B. M. (1994). "Why Compact Fluorescent Lamps Are Not Ubiquitous: Industrial Organization, Incentives, and Social Convention." *Proceedings from the ACEEE Summer Study on Energy Efficiency in Buildings*, Volume 10, 10.77-10.83. American Council for an Energy Efficient Economy, Washington, D.C.

[16] HMG (Heschong Mahone Group). (1997). *Lighting Efficiency Technology Report: Volume III: Market Barriers Report*. P400-98-004VIII, prepared for the California Energy Commission, Sacramento, California. Available URL: http://www.energy.ca.gov/efficiency/ lighting/VOLUME03.PDF.

[17] Gilmore Research Group. (1999). *LightWise: Market Progress Evaluation Report #1*. Report #E99-019, H Mermenet (ed.), Gilmore Research Group, Portland, Oregon.
[18] Decision Sciences Research Associates, Inc., (1998). *Residential Lighting: Market Transformation Study*. Decision Sciences Research Associates, Inc., Pasadena, California.
[19] Gilbride, T. L. (2002). *DOE Sub-CFL Program Exceeds Sales Goal by 200%*. PNNL-SA-37704, Pacific Northwest National Laboratory, Richland, WA.
[20] HMG (Heschong Mahone Group). (1997). *Lighting Efficiency Technology Report: Volume IV: Recommendations Report*. P400-98-004VIV, prepared for the California Energy Commission, Sacramento, California. Available URL: http://www.energy.ca.gov/efficiency/lighting/VOLUME04.PDF.
[21] Gilmore Research Group. (1999). *LightWise: Market Progress Evaluation Report #1: Appendix*. Report #E99-019, Gilmore Research Group, Portland, Oregon.
[22] Dethman, L. (1999). *LightWise, No. 2 (9/99)*. Report #99-038, prepared by Dethman & Associates, Seattle, Washington, with data collected by Gilmore Research Group, Portland, Oregon, for the Northwest Energy Efficiency Alliance, Portland, Oregon.
[23] Calwell, C., Granda, C., Gordon, L. & Ton, M. (1999). *Lighting the Way to Energy Savings: How Can We Transform Residential Lighting Markets? Volume 1: Strategies and Recommendations*. Prepared by Ecos Consulting for the Natural Resources Defense Council, San Francisco, California.
[24] Calwell, C., Granda, C., Gordon, L. & Ton, M. (1999). *Lighting the Way to Energy Savings: How Can We Transform Residential Lighting Markets? Volume 2: Background and Reference*. Prepared by Ecos Consulting for the Natural Resources Defense Council, San Francisco, California.
Saffell, B. F, Imhoff, C. H., Brambley, M. R., Ledbetter, M. R., Jarnagin, R. E., Reilly, R. W., Graybeal, J. W., Weimer, W. C., & Dooley, J. J. (1999). *Transforming Markets for Energy-Efficient Products*. PNNL-SA 32014, Pacific Northwest National Laboratory, Richland, WA.
[25] National Research Council. (2001). *Energy Research at DOE: Was It Worth It? Energy Efficiency and Fossil Energy Research 1978 to 2000*. National Academies Press, Washington, D.C.

[26] Gordon, K. L. & Ledbetter, M. R. (2001). *Technology Procurement Screening Study*. PNNL-13710, Pacific Northwest National Laboratory, Richland, Washington.

[27] Calwell, C., Zugel, J., Banwell, P. & Reed, W. (2002). *2001 – A CFL Odyssey: What Went Right?* PowerPoint Presentation, Ecos Consulting and U.S. EPA, Portland, Oregon.

[28] Calwell, C., Zugel, J., Banwell, P. & Reed, W. (2002). "2001—A CFL Odyssey: What Went Right?" Paper presented at *ACEEE 2002 Summer Study Conference*, Asilomar, California, 6.15-6.27.

[29] ECONorthwest. (2002). *ENERGY STAR® Residential Lighting Program, No. 1: Market Progress Evaluation Report: Final Report.* Report #E02-101, prepared for the Northwest Energy Efficiency Alliance, Portland, Oregon.

[30] Fulbright, V., Jacob, A. & Calwell, C. (2003). *Compact Fluorescent Light Programs Shine through the West Coast Power Crisis*. ER-03-11, prepared by Ecos Consulting for Platts Research and Consulting, Boulder, Colorado.

[31] Seattle City Light. (2003). *Conservation Kit Program Evaluation: Transforming the Residential Use of Compact Fluorescent Lighting*. Evaluation Unit Energy Management Services Division, Seattle, Washington.

[32] Sampson Research. (2004). *Market Effects of BC Hydro's Residential CFL Initiatives: Summary Report.* Prepared for Power Smart Evaluation, Burnaby, British Columbia.

[33] Horowitz, N. (2004). "An Illuminating Summary of California's New Title 24 Residential Lighting Building Code." http://www.energystar.gov/ia/partners/presentations/Noah_Horowitz_Title_24.pdf

[34] Hedman, B., Baggett, S. & Bruchs, D. (2004). *EM&V Report: ENERGY STAR® CFL Program for Small Hardware and Grocery Retailers*. Prepared by Quantec, LLC, Portland, Oregon, for Ecos Consulting, Inc. Portland, Oregon.

[35] ECONorthwest. (2004). *ENERGY STAR® Residential Lighting Program: Market Progress Evaluation Report, No. 2*. Report #E04-130, prepared for the Northwest Energy Efficiency Alliance, Portland, Oregon.

[36] Lighting Research Center. (2003). Increasing Market Acceptance of Compact Fluorescent Lamps (CFLs). Prepared for the U.S. Environmental Protection Agency by Lighting Research Center, Rensselaer Polytechnic Institute, Troy, New York.

[37] RLW Analytics. (2005). California Statewide Residential Lighting and Appliance Efficiency Saturation Study, prepared for San Diego Gas and Electric, Southern California Gas Company, Southern California Edison, and Pacific Gas and Electric by RLW Analytics, Sonoma, CA.
[38] Northwest Energy Efficiency Alliance. (2004). Alliance Accomplishments Toward Its Strategic Mission – 1997 to 2002. Northwest Energy Efficiency Alliance. www.nwalliance.org
[39] Navigant Consulting. (2002). *U.S. Lighting Market Characterization. Volume 1. National Lighting Inventory and Energy Consumption Estimate*. Navigant Consulting, Chicago, for the U.S. Department of Energy, Washington DC.
[40] Energy Information Administration. (1993). "1993 Residential Energy Consumption Survey." U.S. Department of Energy, Washington DC.
[41] Ottman, J. A. (2003). "Green Marketing, Eco-Innovation and Your Customer: Lessons from the Green Graveyard." J Ottman Consulting, Inc. New York.
[42] Kates, Brad, Opinion Dynamics Corp & Steve Bonanno, NSTAR Electric and Gas, "Energy Efficient Lighting in the Residential Market" presented April 2005. Available athttp://www.energystar.gov/ ia/partners/downloads/ meetings/EfficientLightingi nResidentialMarket_Kates_Bonnano.pdf.
[43] California Energy Commission. (2004). "Market Failures, Consumer Preferences, and Transaction Costs in Energy Efficiency Purchase Decisions." CEC-500-2005-020. November 2004. pp. 15-16.
[44] ENERGY STAR Homes Northwest. (2006). Residential Sector Initiative Update, January 2006. http://www.northwestenergystar.com/ files/ 9354RSI_Newsletter_January_2006_.pdf
[45] CEE. (2006). "CEE and NEMA work on 'color' label for CFLs." *CEE Newsletter*, Spring 2006, Consortium for Energy Efficiency http://www.cee1.org/ resrc/news/06-04nl/11_cfl.html

APPENDIX A. A BRIEF HISTORY OF THE DEVELOPMENT OF FLUORESCENT LIGHTING

Incandescent lamps produce light from heat, in this case the heat created by electricity passing through a filament. Fluorescent lamps create light through a two-step process involving chemical reactions and the chemical phenomena known as fluorescence. All fluorescent lamps consist of a tube containing mercury vapor. When an electric charge passes through the mercury vapor, ultraviolet radiation is discharged. The UV strikes chemical phosphors coating the wall of the glass tube; these absorb the UV and "fluoresce" or emit light.

In 1856, Heinrich Geissler developed what was probably the earliest ancestor of the fluorescent lamp when he obtained a bluish glow from a gas sealed in a tube excited by an induction coil. The French physicist Alexandre Becquerel experimented with coating electric discharge tubes with luminescent materials. At the 1893 World's Fair in Chicago, a type of light using fluorescence developed by Nikola Tesla was displayed. In 1894, D. McFarlane Moore created the Moore lamp, a commercial gas discharge lamp using nitrogen and carbon dioxide meant to compete with the incandescent light of his former boss Thomas Edison. In 1901 American Peter Cooper Hewitt demonstrated the low pressure mercury arc lamp, which was very close to the modern design but not quite ready for residential use due to its bluish-green light.

In 1926 Edmund Germer and coworkers in Germany coated the tube with fluorescent powder which converts the ultraviolet light emitted by a rare gas into better spectrally distributed light (also increasing the pressure). General Electric bought the rights to Germer's patent; GE's George Inman and others made several improvements and filed their own patent in April 1936 for the first practical fluorescent lamp, which they began selling in 1938. Several others had patented various less successful versions of a fluorescent lamp, including Edison whose 1896 patent described a fluorescent lamp (never commercially produced) that used x-rays instead of mercury vapor to excite the phosphor. [2,3]

Fluorescent lighting caught on with commercial and industrial building owners due to its cost savings, but the lengthy tubes seemed impractical in residential settings. In the early and mid 1970s, General Electric, GTE Sylvania, Westinghouse, Philips, Lighting Technology Corporation, Spellman Electronics, and others filed patents on designs to shrink fluorescent lights. Most of these efforts proved impractical from a mass manufacturing standpoint, but gradually improvements were achieved, made possible in part by new developments in materials and manufacturing and the shrinking of electronic components. Successful compact fluorescent lamp designs were ultimately put fourth by Philips and Westinghouse and soon by others. All of these designs required substantial investment in new production machinery, [4] a major reason for the high initial price of these lamps [4] and a likely factor in the speed (or lack thereof) with which manufacturers increased production of these new lamps. Some have speculated that Philips' patent also kept prices high for a long time, since manufacturers owed royalties to them until the late 90's or early 2000's whenever they employed rare earth phosphors.

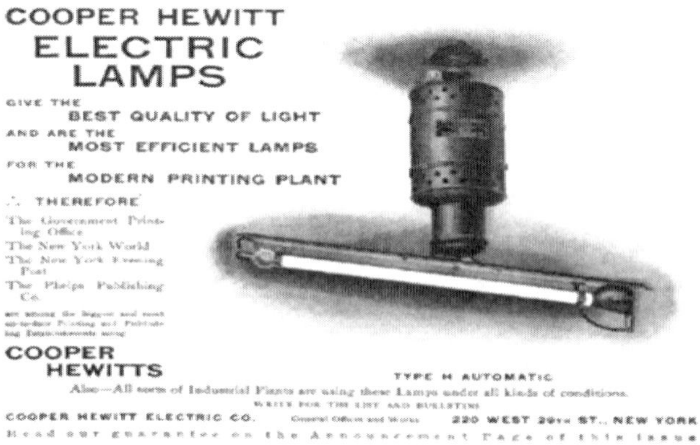

Figure A.1. A drawing of Peter Cooper Hewitt's low-pressure mercury arc lamp taken from the patent application (patent 889,692 issued in 1901), called by some the very first prototype of today's modern fluorescent lamp. [2]

Figure A.2. The "Spiral Lamp" developed by Edward Hammer at GE in 1976. Existing lamp machinery had difficulty making the fragile spiral, and GE felt that new machinery would be too expensive, so they shelved the design. However, spiral lamps appeared on the market in 1995 as other manufacturers decided to see if the design could be competitive. Spiral CFL Smithsonian Institute image #lar2-2d1 [4].

APPENDIX B. A TIMELINE OF EVENTS IN CFL HISTORY

Year	What
1910	Tungsten Light Bulb invented
1933	First Fluorescent tubes installed
1939	Fluorescents made available for wider sale.
1976	Compact Fluorescent "Spiral Lamp" design introduced but GE deems it too fragile to make with existing manufacturing technology.
1976	Hollister introduces magnetic-fluorescent lamp (but not manufactured)
1979	Phillips unveils first electronic ballast CFL
1982	Phillips starts using new rare earth phosphors that emit warmer color of light and increase light output
1985	GE produces first competing product
Late 80s	Utilities start offering CFL bulbs to customers.
1990	April 22, 1990 Earth Day turning point in national consciousness about energy efficiency, climate change, ozone depletion. Three books (*50 Simple Things You Can Do to Save the Earth*, *Consumer Guide to Home Energy Savings*, and *The Green Consumer*) sell millions of copies, recommend CFLs.
1990	Worldwide sales of CFLs = 83 million
1991	CFLs have 1% of US bulb sales volume, 2% of world bulb sales.
1995	Spiral lamps appeared on the market
1997	Philips and TCP introduce dimmable screw-based CFLs
1997	Worldwide sales of CFLs = 356 million
1997	ENERGY STAR Residential Light Fixtures Program started. This brought a benchmark of lighting performance and quality as well as a clearly recognizable national brand to the marketplace.
1998	ENERGY STAR torchieres hit the market and sell a million units by Sept 1999
1998	DOE's b-compact CFL Technology Procurement program started, administered by PNNL Su
1999	The DOE ENERGY STAR screw base CFL program was launched. This continued to help the utilities and regional market transformation groups to rally their marketing strategies.
2000	Beginning of West Coast Energy Crisis
2000	Program for Evaluation and Analysis of Residential Lighting (PEARL) started in 2000 as a watchdog organization in response to complaints received by utility managers about the quality of ENERGY STAR rated products.

Appendix B. (Continued)

Year	What
2001	DOE launches R-CFL Technology Procurement Project to encourage the development of reflector style CFLs that perform well in high heat applications, such as recessed can fixtures.
2001	EU imposes an anti-dumping duty of 66.1 % on CFL-I bulbs, so China turns to US resulting in glut of low-cost CFLs on US market. (EU Regulation 1470/2001)
2001	Rolling blackouts in California prompt massive regional CFL promotions and giveaways
2001	Drought in Northwest prompts regional CFLs programs due to hydro power shortages
2001	Change a Light, Change the World campaign launched with nationwide radio, tv and print advertising of ENERGY STAR CFLs.
2001	New ENERGY STAR requirements including third-party testing and interim life testing.
2001	US CFL sales surge to fourth quarter highs of 2.1% of national lamp market, to 8.5% in California and to 12% in Northwest.
2002	CFL prices are down from $20+ per bulb in 1990 to $9 for a 4-pack without utility subsidy in mass merchandise stores like K-mart, Wal-Mart and Costco around the US.
2003	DOE, American Lighting Association, and Consortium for Energy Efficiency start CFL lighting fixture design competition called "Lighting for Tomorrow"
2005	EU extends the anti-dumping duty from 2001 to include shipments from Philippines, Pakistan, and Vietnam. Apparently some of the Chinese companies affected by the 2001 action had been shipping "kits" of partially manufactured CFLs to these countries for completion and sales. EU Regulation 866/2005
2005	Title 24 (California building code) updates take effect Oct 1,2005. Requires "high efficacy" lights in nearly every room in house. All CFLs must be pin-based, no screw based allowed.

INDEX

A

B

C

D

M

N

O

P

Q

R

U

V

W

X

Z